MILANKOVITCH SEA LEVEL CHANGES, CYCLES AND RESERVOIRS ON CARBONATE PLATFORMS IN GREENHOUSE AND ICE-HOUSE WORLDS

A SHORT COURSE ORGANIZED BY J.F. READ, C. KERANS AND L.J. WEBER

PART 1. OVERVIEW OF CARBONATE PLATFORM SEQUENCES, CYCLE STRATIGRAPHY AND RESERVOIRS IN GREENHOUSE AND ICE-HOUSE WORLDS
by J.F. Read

PART 2. USE OF ONE- AND TWO-DIMENSIONAL CYCLE ANALYSIS IN ESTABLISHING HIGH-FREQUENCY SEQUENCE FRAMEWORKS
by Charles Kerans

PART 3. SEQUENCE STRATIGRAPHY AND RESERVOIR DELINEATION OF THE MIDDLE PENNSYLVANIAN (DESMOINESIAN), PARADOX BASIN AND ANETH FIELD, SOUTHWESTERN USA
by L. James Weber, J.F. (Rick) Sarg and Frank M. Wright

SEPM SHORT COURSE NOTES NO. 35

These SEPM Short Course Notes have received independent peer review. In order to facilitate rapid publication, these notes have not been subjected to the more stringent editorial review required for SEPM Special Publications.

ISBN 1-56576-020-4

Additional copies of this publication may be ordered from SEPM. Send your order to

SEPM
1731 E. 71st Street
Tulsa, Oklahoma 74136-5108
U.S.A.
Tel: (918) 493-3361 ext. 10
Fax: (918) 493-2093

© Copyright 1995 by

SEPM (Society for Sedimentary Geology)
Printed in the United States of America

PART 1

OVERVIEW OF CARBONATE PLATFORM SEQUENCES, CYCLE STRATIGRAPHY AND RESERVOIRS IN GREENHOUSE AND ICE-HOUSE WORLDS

by J.F. Read

OVERVIEW OF CARBONATE PLATFORM SEQUENCES, CYCLE STRATIGRAPHY AND RESERVOIRS IN GREENHOUSE AND ICE-HOUSE WORLDS

J.F. READ

Dept of Geological Sciences, Virginia Tech, Blacksburg, VA 24061, USA

CHAPTER 1

AIMS

This short course/workshop is in 3 parts.

In this part we will examine in general terms how carbonate cycles are generated on carbonate platforms, types of carbonate cycles developed, stacking patterns, margin geometries, degree of disconformity development, and briefly overview any characteristic diagenetic effects. In order to do this, a brief review of controls on carbonate deposition, platform types, and sequence stratigraphy is given. This is followed by discussions of cycle development in greenhouse, transitional and icehouse worlds with brief examination of examples. I would like to emphasize that our understanding of climatic forcing of the cyclic stratigraphic record is still simplistic, and any of the models presented are tentative, and certainly will be modified and refined in the future.

Part 2 (by Charlie Kerans) will examine cycles and one- and two-dimensional stacking patterns, high resolution stratigraphy, and reservoir geometry on Late Permian platforms in the Permian Basin of West Texas, using examples and exercises from Guadalupe Mt. outcrops, and outcrop and core data from a major Permian reservoir. The Late Permian examples typically reflect relatively low amplitude sea level fluctuations, following collapse of the Permo-Carboniferous ice-sheets.

Part 3 (by Jim Weber, Rick Sarg and Frank Wright) will examine reservoirs formed in an ice-house world during the major Carboniferous glaciation of Gondwana, using the Middle Pennsylvanian carbonates of the Giant Aneth oil field, Paradox Basin, Utah. Again, outcrop and subsurface data are used to develop a regional sequence stratigraphic framework, and the controls on facies development and reservoir quality of these stratified reservoirs examined.

CONTROLS ON CARBONATE DEPOSITION

Excellent reviews are given in Vail et al. (1991) and Schlager (1992).

Subsidence:

Thermal and tectonic subsidence (driving subsidence) related to stretching and cooling of the lithosphere creates the initial space for sedimentation in rift basins, passive margins and cratonic basins (Fig. 1-1A). In rifted regions, small platforms may develop above rift blocks which undergo rotation as stretching proceeds. On passive margins, thermotectonic subsidence tends to decrease exponentially with age (Fig. 1-1A). On most passive margins, subsidence rates commonly are a few cm to 15 cm/k.y. Subsidence rates commonly are much faster in rift and foreland basins (tens of cms/k.y.).

The isostatic loading associated with newly emplaced water and sediment in sedimentary basins creates additional space due to isostatic flexural subsidence. Assuming local isostasy and neglecting flexure, a widespread transgression of an emergent shelf could cause roughly an additional 40% subsidence just due to water load. Similarly, transgressing a previously emergent shelf and filling this with sediment could cause a total sediment pile that is 2 to 3 times the thickness of the initial depth of water. Conversely, eroding sediment results in uplift due to unloading. Isostatic response times appear less than a few thousand years. To find the amount of driving subsidence involved in deposition of a basin fill, one needs to remove the subsidence due to loading by progressive backstripping the sediment using a geohistory plot or burial history curve (total subsidence vs time) (Van Hinte, 1978) (Fig. 1-1A). In general, long-term subsidence rates are less than potential

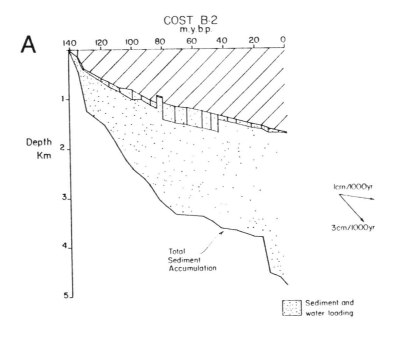

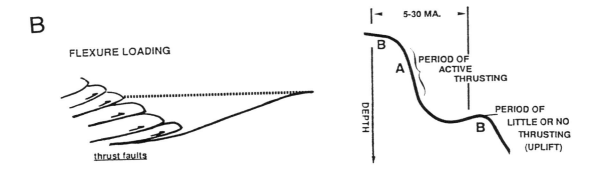

Fig. 1-1. Subsidence history plots. A. Example of subsidence history plot from a passive margin (COST B-2 WELL). Lower curve is total subsidence whereas upper curve (vertical hachuring) is backstripped subsidence or thermo-tectonic subsidence after sediment and water loading have been removed given the estimates of paleobathymetry (from Watts, 1981).
B. Schematic diagrams illustrating subsidence due to flexural loading by thrust sheets (left) and subsidence history plot on the right (from Vail et al., 1991).

sedimentation rates on most carbonate platforms and will not cause drowning by themselves.

In foreland basins, loading due to thrust emplacement is a major cause of rapid subsidence (Fig. 1-1B). Active foreland basins may have craton-attached carbonate platforms, or during major phases of tectonic quiescence and overfilling, carbonate platforms also may extend out from the orogenic highlands onto the deeper, sediment starved craton . Foreland basins can exhibit complex differential subsidence patterns, in part due to complex pre-existing basement structures (Dorobek, 1995). Foreland basins may develop stratigraphies that are different from passive margins or cratonic basins (Posamentier and Allen, 1993; Flemings and Jordan, 1991). This is because craton-attached carbonate platforms extending into foreland basins are receiving most of the siliciclastic sediment from the side with the highest subsidence rate and there may be periodic upwarping associated with a peripheral bulge during thrusting. During thrusting, there is ponding of coarse siliciclastics on the tectonically active margin and the basin undergoes maximum deepening. There may be apparent offlap along the cratonic margin synchronous with uplift on the peripheral bulge. Thus the regressive-transgressive events forming the sequence boundary may be out of phase on the distal vs proximal margins. With cessation of thrusting, erosion of the tectonic highlands coupled with decreased load-induced subsidence causes rapid progradation of coarse clastics out across the basin and onlap onto the cratonic margin.

Foreland basins also will show a proximal zone near to the tectonically active basin margin in which subsidence always exceeds eustatic sea level fall, and hence will contain only conformable sections. In contrast, further out from the tectonically active margin, subsidence rates will sometimes be exceeded by eustatic sea level fall, and so will develop unconformities in the section.

Sea-Level Change:

Global sea level changes with time due to 1. changes in the volume of the ocean basins, in part due to heat flow through mid-ocean ridges and 2. due to global ice volume.

First order cycles 200 to 300 m.y. long commonly relate to plate reorganization, starting with breakup of supercontinents, opening of ocean basins, and ultimate closure (Wilson cycle) (Vail et al., 1977). These cause the long term cratonic onlap and offlap observed (e.g., Cambrian to Permian.

Second order cycles 10 to 50 m.y. long (Fig. 1-2) that are driven by tectonics and change in ocean basin volumes and to a lesser extent by ice-volume, form the widespread major depositional sequences we commonly see. Thicknesses commonly are hundreds to a few thousand meters.

Superimposed on these are third order cycles 0.5 to 5 m.y. long (Fig. 1-2) which form the smaller scale depositional sequences that we observe. Sea level changes commonly are 50 m or less, and rates are a few cms/k.y. Probably eustatic in origin and related to ice volume.

High frequency, climatically driven sea level cycles from less than 20 k.y. to 400 k.y. are superimposed on these longer term cycles, their amplitude depending on global ice volume during greenhouse and ice-house times (Figs. 1-3 to 5) (Fischer, 1964; 1982; Goldhammer et al., 1990; Mitchum and Van Wagoner, 1991) and it is these that cause rapid flooding of platforms. These climate changes result from cyclic changes in the shape of the earth's orbit, and changes in the tilt and wobble of the axis, and have been termed Milankovitch cycles (Fig. 1-4 and 5). These have periods of roughly:

A. approximately 100 and 400 k.y. (short- and long term eccentricity); the dominant cycles during times of maximum glaciation (Fig. 1-5). Although eccentricity modulates the precessional signal, just why eccentricity is so strong during glacial times is not clear, as the amount of direct insolation available at the eccentricity rhythm is very small

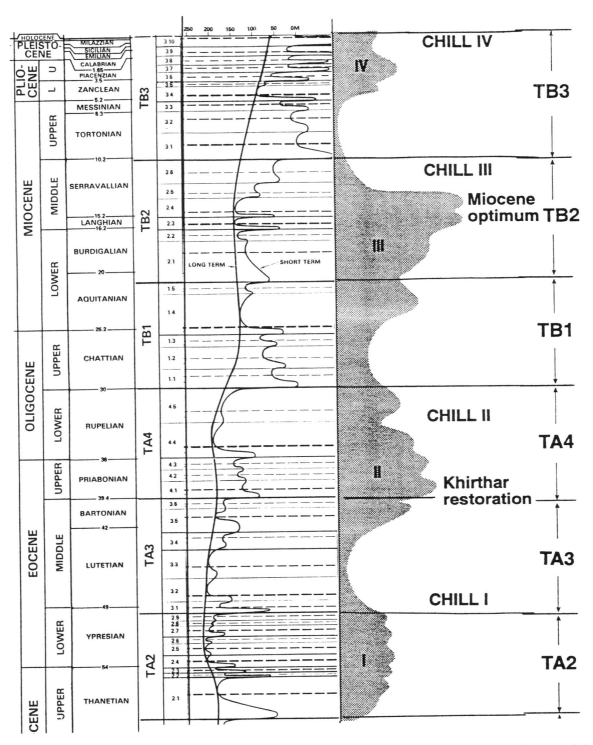

Fig. 1-2. Sea level cycle chart illustrating the roughly 10 m.y. cycles and superimposed 1 to 5 m.y. cycles (modified from Haq et al., 1987). Stipple plot on right hand side shows major sea level cycles from southern Australia, with highstands plotted toward right (mirror image of Haq et al. curve), with the major chilling events shown. Note that many of the major lowstands and early transgressions correspond to major cool phases (from McGowran et al., 1994).

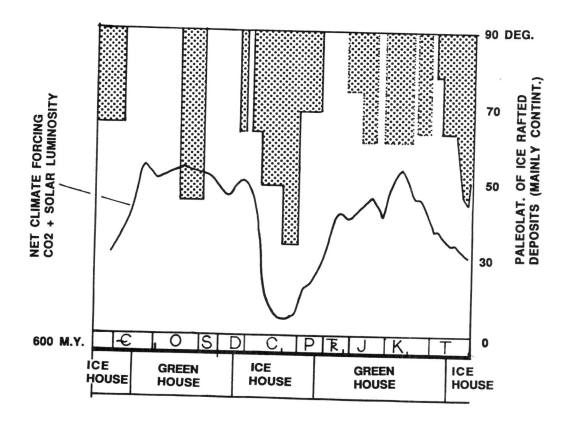

Fig. 1-3. Relationship between ice house and green house conditions, eustasy and global CO2. Upper part of graph shows paleolatitudes of ice-rafted continental glacial deposits (stippled with bold outlined bars) and marine ice rafted deposits (dotted outlined bars), modified from Frakes and Francis (1988), plotted against geological time, as well graph showing net forcing of climate due to changes in CO2, their effect on radiative forcing, and long term increase in solar luminosity (modified from Crowley and Baum, 1992 and Berner, 1991). Times of generalized ice house vs. green house conditions are shown on the time axis, and are modified from Fischer (1982).

(Ruddiman and Wright, 1987). It has been suggested to be a beat frequency produced by the 19 to 23 k.y. cycles, or more likely a non-linear responses in the climate system which amplifies the effects of direct insolation to give the observed ice-sheet response.

B. approximately 40 k.y. (obliquity or tilt). Obliquity apparently may be evident during greenhouse times, causing a bundling of precessional cycles. It may be more important during transitional and ice-house times. For example, it is the strongest signal in the pre-700 k.y. Pleistocene record and reflects insolation forcing at high latitudes (Fig. 1-5) (Ruddiman and Wright, 1987).

C. 19 to 23 k.y. (precession or wobble). Precessional carbonate cycles appear to be the dominant sedimentary cycles during times when the earth is relatively ice-free (Fig. 1-5). Precessional and obliquity cycles in the Early Paleozoic may be shorter than those today (roughly 17 and 30 k.y. respectively) (Berger et al., 1989).

D. Sub-Milankovitch cycles 10 k.y. or less. Cycles around 12 k.y. or less have been recognized in the stratigraphic record (Olsen et al., 1993; Bond et al., 1991; Kominz et al., 1991; Mundil et al., in press); these might be precession beats (constructive interference of precessional quasi-periods)(Kominz et al., 1991). Also, since precession would be expected to influence alternate hemispheres of the earth every 10 k.y. or so, these might be "precessional half-cycles". Higher frequency sub-Milankovitch cycles also are evident in the stratigraphic record, down to the strong 2.5 k.y. cycle evident in ice-cores and slope facies on carbonate platforms (see Elrick et al., 1990 for references).

Note that the above periods are the dominant ones observed. A glance at time-series from the Pleistocene (Fig. 1-5) illustrates the point that sea level changes are unlikely to be simple 20, 40, 100 and 400 k.y. fluctuations, because a) there are numerous quasi-periods within the precession, obliquity and eccentricity bands, b) the orbital forcing-climate-glaciation-sea level response is complex and non-linear c) shallow carbonate platforms rarely preserve all the sea level fluctuations and d) other cycles such as autocycles and sub-Milankovitch cycles are likely to be present in the stratigraphy. Thus it is simplistic to look for simple 5 to 1 or 40 to 1 bundles as tests of Milankovitch forcing. However, there is little doubt that Milankovitch orbital forcing of climate has had a profound influence on the stratigraphic record since the birth of the solar system (Archean to Recent).

During times of continental glaciation (Pleistocene or Pennsylvanian and Early Permian), 100 to 400 k.y. sea level changes are large (up to 100 m). Forty k.y. cycles also may be important during these glacial times, especially when global ice volume is reduced (Fischer, 1986; Ruddiman et al., 1986). During times of continental glaciation, these sea level changes have a rapid rise (deglaciation) and gradual fall (glaciation). These glacio-eustatic sea level changes cause rapid transgressions of meters/k.y. that far exceed most sedimentation rates. At the other extreme, during greenhouse times when the earth had little ice (Late Cambrian, Middle Devonian, Middle Triassic) sea level fluctuations are small, commonly less than 10 m, and may be dominated by precessional cycles (20 k.y. or less) and possibly low amplitude 40(?), 100 and 400 k.y. cycles which generate bundles of cycles (Goldhammer et al., 1990; Koerschner and Read, 1989; Wright, 1992). During greenhouse times with little ice, these high frequency sea level changes could be driven in part by changes in lake volumes (Jacobs and Sahagian, 1993).

Sea Level Influence on Climate:

In situations where sea level is driven by glacio-eustasy, then during deposition of a single depositional sequence (say 1 to 10 m.y. duration) low stands and subsequent transgressions may be associated with cooler climate and more extensive ice-sheets than during subsequent highstand. This situation is suggested by Tertiary data (Pomar, 1991; McGowran et al., 1994) (Fig. 1-2) Milankovitch driven sea level changes that are superimposed on the long term low-stand and early transgression, therefore might be higher amplitude than during long term highstand.

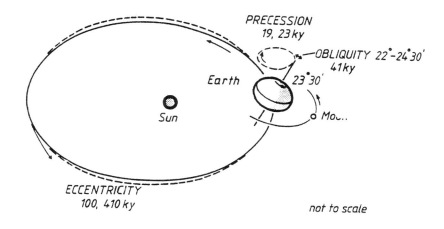

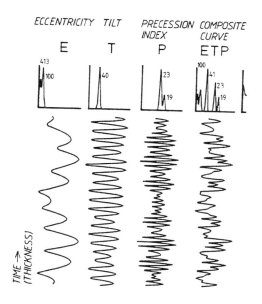

Fig. 1-4. Top: Diagram illustrating Milankovitch orbital forcing of climate. Bottom: Diagram showing how 100 and 400 k.y. eccentricity, 41 k.y. obliquity and 19 to 23 precessional cycles interact to develop the climate change signal and hence sea level change (composite curve). Diagrams from Einsele and Ricken, 1994.

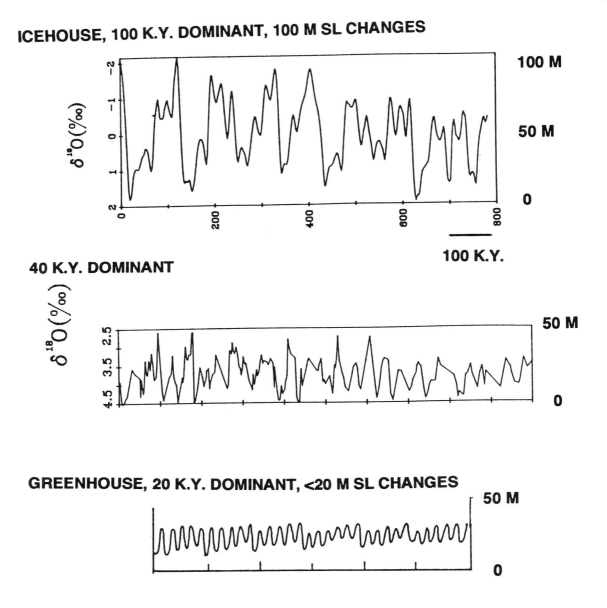

Fig. 1-5. Examples of inferred ice-house vs greenhouse high frequency sea level curves roughly plotted at same scale. Curves become younger toward the left. Small bar beneath horizontal axis is 100 k.y. Top: ice-house oxygen isotope curve which likely is a proxy for sea level, shown in meters on scale at right. Curve shows high amplitude signal with dominance of 100 k.y. cycles and is typical of last 700 k.y. of the Late Pleistocene. Middle: Oxygen isotope curve for earlier Pleistocene, showing dominance of 40 k.y. cycles, and lower amplitude. Both from Ruddiman and Wright (1987). Bottom: inferred greenhouse sea level curve inferred to be dominated by low amplitude precessional sea level changes.

Similarly, during long-term highstand, when globally warmer conditions reduce global ice volume, Milankovitch sea level changes are likely to be much lower. This could result in depositional sequences showing evidence of increasing water temperatures, and decreasing magnitude of Milankovitch eustasy from the long term lowstand to the highstand. A consequence of cooler conditions during low-stand could be decreased sedimentation rate on the platform, which would promote transgression and incipient drowning of the platform.

Long term sea level change can strongly influence rainfall and hence climate, especially in the case of Paleozoic and Mesozoic epeiric seas. During long term, late highstand and lowstand when the seas migrate to the edge of the platforms and leave much of the continental interior emergent, conditions are likely to be relatively arid; this aridity might be amplified by lack of land plants in the Precambrian to Silurian(?), in view of the immense moisture transfer into the atmosphere by evapo-transpiration by vegetation today. Aridity would tend to promote dolomitization of highstand carbonates in coastal areas and development of caliches in inland areas. During long-term transgression, development of extensive shallow seas over the continental interior could promote increased rainfall. This wetter climate might cause karstic features to be superimposed on semi-arid caliche deposits. Wetter climate would also promote increased vertical stratification of the sea, and development of relatively shallow (tens of meters) anoxic bottom waters and associated black shales or carbonates. This would suggest a common motif in carbonate depositional sequences of more humid conditions/facies during transgressive systems tracts, and more arid climate/facies during late highstand-lowstand systems tracts.

High frequency, Milankovitch driven climate changes (besides making polar ice) may cause 10 k.y. to 400 k.y. cyclical alternation of climate (in terms of temperature and rainfall) in carbonate-forming low latitude areas (Matthews and Perlmutter, 1994). Such cyclic climate changes may cause climates to alternate from hot to cool, and from arid to humid, depending on the platform location (Fig. 1-6). For example, in the tropics during a climatic maximum (warm phase) the humid belt may be relatively broad. However, with cooling and contraction of the Hadley cell, the humid belt may shrink leaving the area more arid.

Similarly, outside the tropics during the climatic optimum, descending air associated with high pressure would cause widespread aridity. However, with the climatic minimum and cooling, shrinkage of the Hadley cell would allow temperate, more humid conditions to develop in the formerly desert area.

These Milankovitch driven climate changes can cause cyclic alternation of terrigenous input to the platform, from little clastic material during arid times to abundant clastic influx during humid times, as well as changes in amount of diagenesis caused by fresh water. One might expect the climatic changes to be most intense during ice-house times and continental glaciation, when pole-to-equator temperature gradients are at a maximum. Conversely, during greenhouse times, such Milankovitch changes presumably would be less intense. Coastal rainfall effects similar to those discussed for longer term transgression-regression could complicate this scenario.

Rate of Sediment Accumulation:

Rates of carbonate sedimentation in tropical settings tend to be strongly dependent on light, and hence water depth because of the biotas (corals, green algae, sea grasses, termed chlorozoan association). Waters in open ocean settings are well lit down to 100m, but in basins such as the Persian Gulf, the photic zone may be as little as 20 to 30 m depth. This is important to remember when dealing with basins receiving fine material. There is a rapid drop off in growth rate below 10 m (Fig. 1-7) hence tropical faunal assemblages tend to form reef-rimmed platforms (Schlager, 1992). These reef builders also tend to be limited to near normal

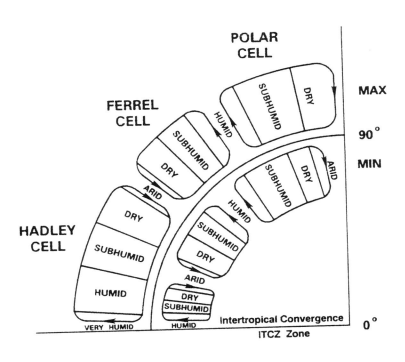

Fig. 1-6. Highly simplified atmospheric circulation model showing how climates may be greatly modified during climatic maxima and climatic minima. From Matthews and Perlmutter, 1994.

salinities, and to waters warmer than 20°C winter temperature (roughly between 30 degrees latitude). Tropical carbonate sedimentation rates generally easily exceed subsidence rates (Fig. 1-8), however, only reef assemblages are able to keep pace with high amplitude sea level rises of several meters/k.y. (Fig. 1-8).

In contrast, cool and cold water carbonates do not show such a light dependence because assemblages are dominated by bryozoans, mollusks and forams (termed bryomol or foramol association). Consequently, although temperate water carbonates do not show high shallow water production rates, they show little decrease in sedimentation rates into deeper water, thus these tend to form gently sloping ramps on prograding seaward thickening sediment wedges (Schlager, 1992; Lees, 1975) (Fig. 1-7). Shallow water settings on such temperate shelves are mainly sites of sediment starvation, extensive wave abrasion and sediment reworking (Collins, 1988). Such cooler water carbonates typically occur at higher latitudes. However, cool water carbonates (with updip, shallow, warm water facies) also may occur in low latitude areas that are sites of upwelling, or thermal stratification (James, 1994; Martindale and Boreen, 1994).

Clastic poisoning due to turbidity reducing light or suffocating filter feeders is important in reducing carbonate production rates toward sites of fine clastic influx and may be a major influence on intrashelf basin formation.

Lowered salinities associated with clastic influx also likely will reduce production rates. In contrast, higher salinities on arid platforms reduces biotas to all but the most tolerant types. On interiors of huge shallow water Paleozoic platforms, distance from the ocean and its source of calcium ions also may keep production rates low, resulting in cratonic deeper water carbonate blankets and intrashelf basins, which are located behind extensive shallow water platforms.

Change in Biota Through Time:

Carbonate producing organisms have changed significantly through time (James, 1983). For example, reef builders similar to modern coral-coralline algal assemblages of the Late Cenozoic are best developed in the Siluro-Devonian (stromatoporoids), and perhaps parts of the Triassic-Jurassic (corals, sponges, stromatoporoids) and Mid-Cretaceous (rudist clams). At other times, shelf edges were occupied by delicate branching and encrusting organisms, such as the Permo-Triassic Tubiphytes, and the Cambrian skeletal algae. Similarly, forams only become abundant in the Late Paleozoic, and continue to the present as major sediment producers forming banks and shelf sands. Bryozoans were important mound builders in the Paleozoic even in relatively tropical settings, whereas today they are most widespread on cool water, deeper continental shelves. Crinoids were major sediment producers on Paleozoic platforms, but are relatively rare after this. Such marked biotic changes through time need to be taken into account when applying any carbonate models to the subsurface.

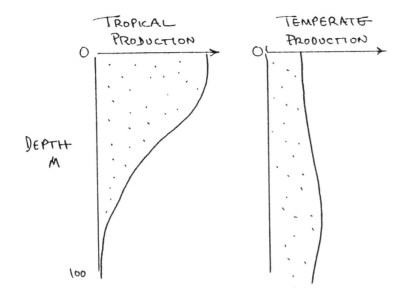

Fig. 1-7. Sediment production/accumulation rates. Left: Production vs depth curve for tropical area, showing maximum production in shallow water decreasing rapidly (below 10 to 40 m) to low values, because of the light dependence of the biota. Right: Production vs depth curve for temperate water carbonates showing the much slower decrease in production into deeper water, reflecting the dominance of organisms (e.g. bryozoans, crinoids) that are not light dependent.

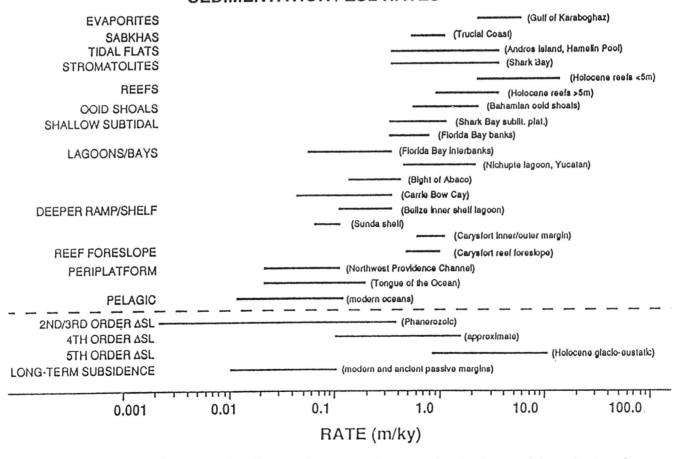

Fig. 1-8. Table showing plot of sedimentation rates (horizontal axis, log scale) against various environments, as well as the ranges for various eustatic changes and long term subsidence. Based on a table in Schlager (1981) and modified by D. Osleger.

CHAPTER 2

BASIC TYPES OF CARBONATE PLATFORMS

For a more detailed review of carbonate platform types and evolution see Read (1985) and Handford and Loucks (1993).

Standard Facies Belts of Carbonate Platforms

- Tidal Flat Facies: Facies arranged in cyclic, upward shallowing units 1-10 m thick. In humid zones facies are subtidal-intertidal burrowed limestone with supratidal cryptalgal laminites, and inland freshwater algal marsh deposits, coal, or siliciclastics. In arid zones facies are commonly dolomitized, burrowed to non-burrowed pellet muds and cryptalgal heads, overlain by abundant intertidal cryptalgal laminite sheets, supratidal evaporites, or eolian-fluvial clastics. RESERVOIR FACIES commonly in dolomitized subtidal units.
- Lagoonal Facies: Mainly bedded pellet limestone or lime mudstone, or cherty, burrowed skeletal packstone to mudstone, with local biostromes of colonial metazoans. Minor, thin interbeds of peritidal fenestral or cryptalgal carbonates reflecting periods of shallowing of lagoon to tide levels. RESERVOIR FACIES if leached.
- Shoal-water complex of banks, reefs, and ooid/pellet shoals: Occur as shallow-ramp skeletal banks or lime-sand shoals, or shelf-edge skeletal reefs and skeletal or oolitic-peloidal sands. On ramps, they pass gradually downslope into deep-ramp facies. On steeply sloping rimmed shelves, they pass downslope into foreslope and slope deposits. Note the term bank refers to relatively in place, biostromal to mound-like skeletal accumulations lacking a rigid internal skeletal frame; a reef is sheet-like to mound-like in-place skeletal accumulations with a rigid internal skeletal frame. Reefs and banks thus are defined on presence or absence of frame. RESERVOIR FACIES commonly in reefs and banks associated with primary porosity and in dolomitized equivalents.
- Deep shelf and ramp facies: Cherty, nodular bedded, skeletal packstone or wackestone, with abundant whole fossils, diverse open-marine biotas, and upward-fining, storm-generated beds. Water depths 10 to 40 m and largely below fair-weather wave base, but commonly above storm wave base. RESERVOIR FACIES commonly occur in dolomitized and leached updip portions of these muddy carbonates.
- Slope and basin facies: Adjacent to steeply sloping platforms, foreslope and slope deposits have abundant breccias and turbidites interbedded with periplatform lime and terrigenous muds. Adjacent to ramps, slope and basin deposits are thin-bedded, periplatform lime and terrigenous muds that generally have few sediment-gravity flow deposits. Basinal deposits in Paleozoic rocks commonly are shale, with carbonate-content increasing toward the platform. Basinal deposits in Mesozoic and Cenozoic rocks may be shale or pelagic limestone. Slope and basin floor may be anoxic and lacking benthic organisms; thus, deposits will be laminated and non-burrowed. Where slope and basin waters are oxic, deposits may be burrowed and fossiliferous. RESERVOIR FACIES may occur in breccias and lime sands.

Carbonate Ramps

Ramps (Fig. 2-1) have gentle slopes (generally less than 1°) on which shallow wave-agitated facies of the nearshore zone (ooid-peloid shoals or skeletal banks) pass downslope (without marked break in slope) into low energy, deeper ramp muds. They differ from rimmed shelves in that continuous reef trends generally are absent, high-energy lime sands are located near the shoreline (rather than on a shelf edge), and deeper water breccias (if present) generally lack clasts of shallow shelf-edge facies. The high energy shoals on ramps can be either fringing or barrier banks or fringe or barrier ooid-peloid shoals.

- Homoclinal ramps have relatively uniform, gentle slopes (1 to a few meters/km or a fraction of a degree) into the basin (Fig. 2-1) and lack a sharp break in slope (Fig. 2-2). Ramps show gentle, seaward dipping reflectors on seismic cross-sections, however, on some ramps prograding into deep basins, slopes can be many degrees.
- Distally steepened ramps have the shoal-water complex well back on platform separated from the slope by a broad deep ramp; the slope facies contain abundant slumps, breccias, and allochthonous lime sands but deep-water breccias lack clasts of shallow-platform sands or reefs, and only contain clasts of deep-ramp or slope facies (Fig. 2-3,2-4).
- Low-energy, distally steepened ramps have widespread deep-ramp mud blankets seaward of the shoal-water complex.
- High-energy, distally steepened ramps only occur on swell dominated continental margins (e.g. southern and southwestern Australia). They have broad lime-sand blankets over much of the deep ramp, with muds (and slope breccias and turbidites) being restricted to the slope and basin margin. Shoal-water complexes on homoclinal and low-energy, distally steepened ramps include skeletal banks or ooid-pellet sand shoals; these may be either fringing or barrier complexes. High-energy, distally steepened ramps have wide beach-dune complexes, and extensive thin shelf sand blankets extending to depths of over 100 m.

Rimmed carbonate shelves

Rimmed shelves (Ginsburg and James, 1974) are shallow reef-rimmed, flat-topped platforms whose outer wave-agitated rim slopes steeply into the basin; slopes commonly are a few degrees to 60° or more) (Fig. 2-5). They have a semicontinuous to continuous reefal rim or barrier along the shelf edge, backed by skeletal or oolitic sands. Foreslope sands and muds pass downslope into breccias and turbidites. Seismic expression of reefs is shown in Fig. 2-6.

- Depositional or accretionary rimmed shelves show both up-building and out-building; they generally lack high marginal escarpments; and shelf edge and foreslope/slope facies intertongue (rather than abut). Slopes generally are steep, and flatten with depth, but in some youthful rimmed platforms, slopes may be ramp-like because the platforms have not had time to build significant bank-to-basin relief, or high sedimentation rates in the basin keep it almost filled.
- Bypass margins of rimmed shelves occur in areas of rapid upbuilding where shallow water sedimentation keeps pace with sea-level rise. Bypassing may be associated with a marginal escarpment and/or a gullied bypass slope.
- Erosional margins commonly are characterized by high, steep erosional escarpments up to 4 km relief. Reefal carbonates rim the platform, and are exposed on the upper few hundred meters of the upper escarpment. Downslope, due to erosional retreat of the escarpment by mechanical defacement, bedded, cyclic lagoonal, and peritidal beds are exposed along the escarpment. Erosional margins are relatively common in the geologic record, both in outcrop and in seismic sections (Fig. 2-7).

Many rimmed shelves have inshore or intrashelf basins lying behind the shallow carbonate rim. The basins commonly pass landward into coastal siliciclastics and to seaward into the shallow carbonate rim by way of a gently sloping ramp. Intrashelf basins have water depths of a few tens of meters, and lie below fair-weather wave base but parts may be above storm wave base. Sediment fills are shale with thin beds of quartz- and lime silt, intraformational conglomerate, glauconite, and radial-ooid packstone in storm-generated, upward-coarsening, and upward-fining units. Subwave base fills may be euxinic to dysaerobic organic rich limestone and shale that form excellent source beds.

Isolated platforms

These mainly are reef rimmed, are separated from continental shelves by deep water, and commonly are tens to hundreds of kilometers wide. On continental margins they are located above rifted continental or transitional crust. In oceanic settings, they overlie volcanic sea mounts (modern day atolls). Platform to basin transitions resemble those of rimmed shelves (Fig. 2-5). Interiors of reef-rimmed platforms may be dominated by skeletal limestone, where interiors are relatively deep (up to 20 m). In contrast, where platforms are shallow and flat-topped, interior facies may be dominated by cyclic nonskeletal peloidal sands and muds, and platform margins are shoals and eolian islands of ooid grainstone with subordinate reefs. One of the major differences between isolated platforms and other types is that margins may be windward (sediment is swept onto the platform), leeward (sediment is swept off the platform, inhibiting reef growth and forming wide deep water mud aprons. Others are tide-dominated and have spectacular ooid shoals. Some large platforms have formed by coalescence of smaller platforms following infilling of intervening deep water passages.

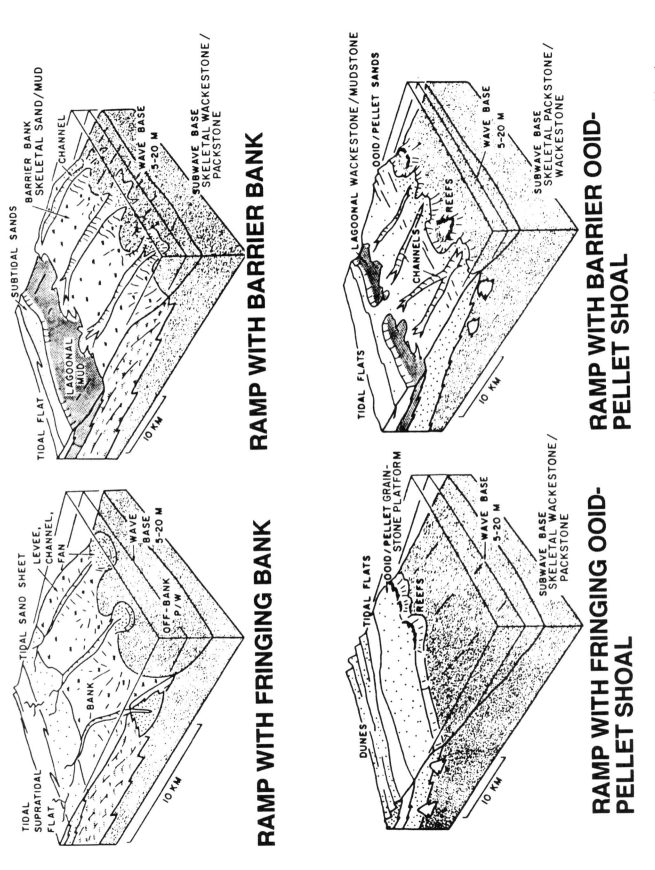

Fig. 2-1. Block diagrams illustrating carbonate ramp facies associated with fringing and barrier banks and fringing and barrier ooid-peloid complexes

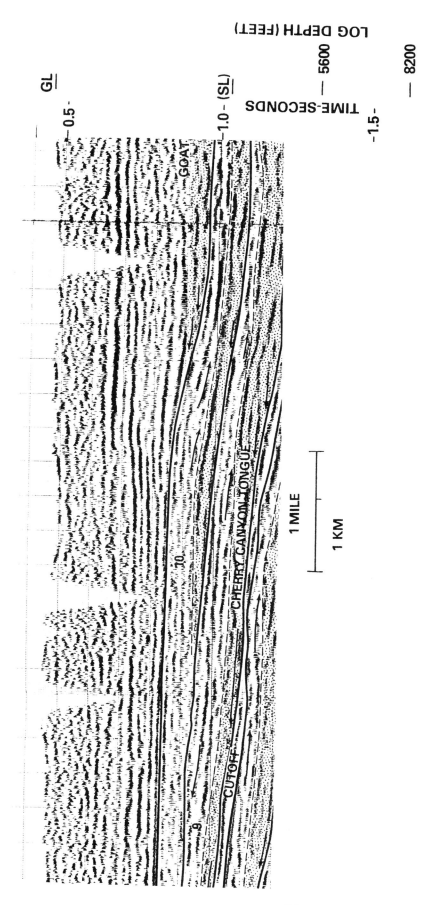

Fig. 2-2. Seismic expression of Late Permian Grayburg ramp, Permian Basin, Texas. Slopes on some of these Permian ramps are steeper than many ramps. Stippled pattern indicates low stand mixed carbonate and siliciclastic sands. From Sarg and Lehmann, 1986.

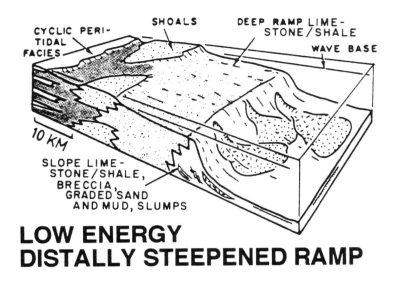

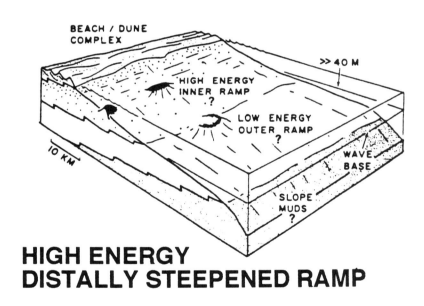

Fig. 2-3. Distally steepened ramp models for low energy margins and high energy, swell dominated (open ocean) margins.

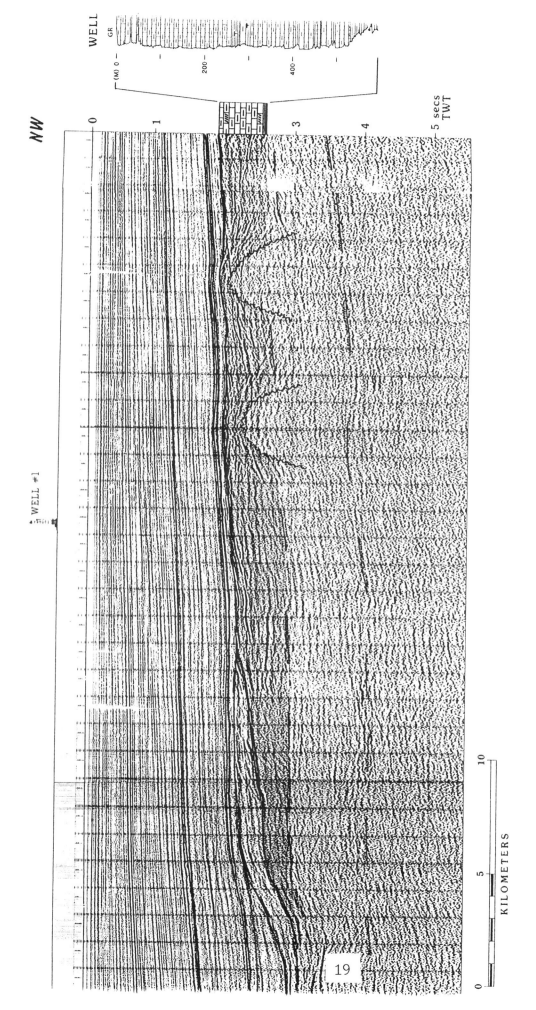

Fig. 2-4. Seismic expression of possible distally steepened ramp, Jurassic of the Grand Banks, Canada. Updip intrusive structures are salt. From Hubbard, Pape and Roberts, 1986.

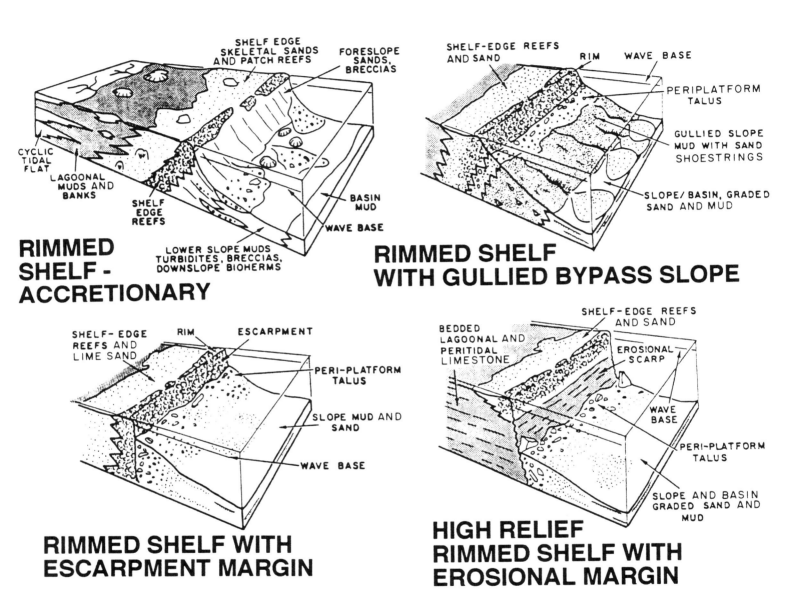

Fig. 2-5. Block diagrams of various types of rimmed shelves arranged in terms of increasing relief - accretionary rimmed shelf to rimmed shelf with gullied bypass slope to escarpment margin to high relief erosional margin.

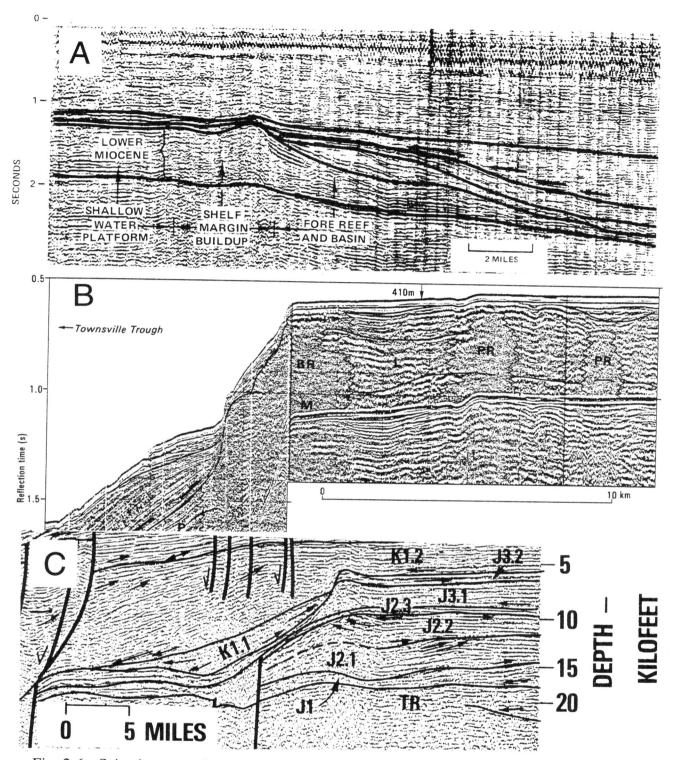

Fig. 2-6. Seismic expression of rimmed shelves. A. Gulf of Papua, Lower Miocene. Margin appears to vertically aggrading and lacks any evidence of an escarpment or erosional face. Bubb and Hatfield, 1977. B. Drowned Neogene reef-rimmed shelf showing evidence of low escarpment and backstepping prior to drowning. Marion Plateau, NE Australia. M is multiple. RR is reef rim, PR is patch reef, L is lagoonal facies. From Davies, 1989. C. High relief Jurassic reef rimmed margin, offshore West Africa which evolved from more ramp-like margin. Shelf margin shows several thousand feet of relief between the platform top and the basin floor, and likely is an erosional bypass margin. Todd and Mitchum, 1977.

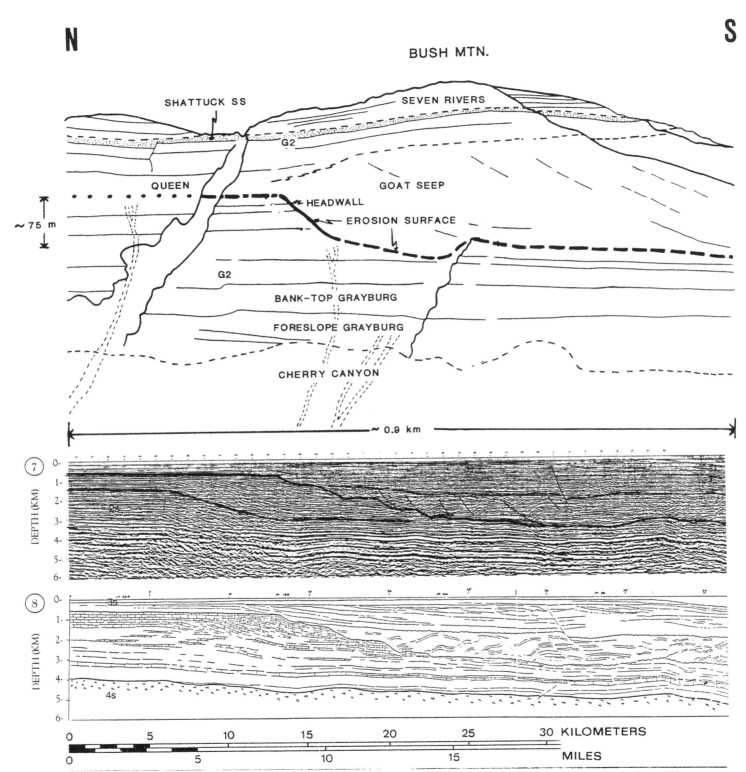

Fig. 2-7. Top: Outcrop expression of erosional margin, Grayburg Formation, Late Permian, West Texas. Franseen, 1989.
Bottom: Seismic expression of erosional carbonate margin, Tertiary of offshore Guyana showing large scale slumps derived from the carbonate platform transported into muddy basin facies. Jankowsky and Schlapak, 1983.

CHAPTER 3

DEPOSITIONAL SEQUENCES, PARASEQUENCES AND SYSTEMS TRACTS

Good summaries of definitions of sequence stratigraphy are given in Vail (1987), Van Wagoner et al. (1988), Sarg (1988), Schlager, 1992, and Handford and Loucks (1994). The sequence stratigraphic approach was developed mainly from seismic sections in which sequences and systems tracts were defined largely on stratal geometries (e.g. coastal onlap, offlap, downlap) (Fig. 3-1A). In contrast, much depositional sequence stratigraphy at the reservoir scale is done using well logs, cores or outcrop sections, and thus depends on a totally different type of data set. Integrating these two approaches is the challenge facing today's explorationist.

The sequence stratigraphic approach allows us to break up a basin's stratigraphy into genetically related packages termed depositional sequences. Idealized sequence stratigraphic models for rimmed shelves and ramps are shown in Figures 3-1 and 2. The depositional sequences are bounded updip by unconformities and downdip by correlative conformities. Sequence stratigraphic units result from the interaction of

1. rates of subsidence,
2. rate of eustatic sea level change and
3. sedimentation rate.

The direction of long-term movement of the shoreline - is a function of sedimentation rate vs. rate of creation of space on the shelf (accommodation).
transgression: is landward movement of the shoreline -
regression: is seaward movement of the shoreline -
Accommodation is related to rate of subsidence and eustatic sea level change. Although eustatic sea level fluctuations are important in formation of depositional sequences, tectonics and varying sediment supply also are important and may be difficult to separate without high resolution biostratigraphic data on a global scale. Consequently, relative sea level curves which are the sum of tectonic subsidence and eustatic sea level change can be employed where global eustatic curves are poorly documented.

Major depositional sequences (termed 2nd order) are 10 to 50 m.y. duration, and commonly contain minor depositional sequences 0.5 to 5 m.y. duration (termed 3rd order). They commonly are resolvable at the seismic scale with the resolution decreasing as the sequences become thinner, and higher frequency.

Sequence-bounding unconformities are initiated at times when the rate of sea level fall exceeds the rate of subsidence. Thus, because subsidence rates increase seawards on most platforms, the unconformities pass downdip into correlative conformities.

Systems tracts which consist of all the facies (subaerial to tidal flat to shelf to slope and basin) deposited during either low stand, transgression or highstand, make up the depositional sequences. These are termed low-stand (LST), transgressive (TST) and highstand systems tracts (HST). The transgressive surface (ts) or flooding surface separates the LST from the TST. The maximum flooding surface (mfs) separates the TST from the HST.

Most systems tracts are themselves composed of small scale shallowing upward cycles from a meter to 10 m or more (parasequences) which are bounded by minor flooding surfaces (Figs. 3-1B and 3-2). Parasequences are the familiar small scale cycles that we see in the field. On the inner platform, they consist of shallow lime sands or muds capped by tidal flat facies and exposure surfaces. On the shelf edge or shoal complexes of ramps, they may be poorly cyclic to non-cyclic, and grainstone/reef/bank dominated. In slightly deeper water, parasequences show coarsening upward from muds and interbedded muds and sands up into grainy caps.

Most parasequences relate in part to high

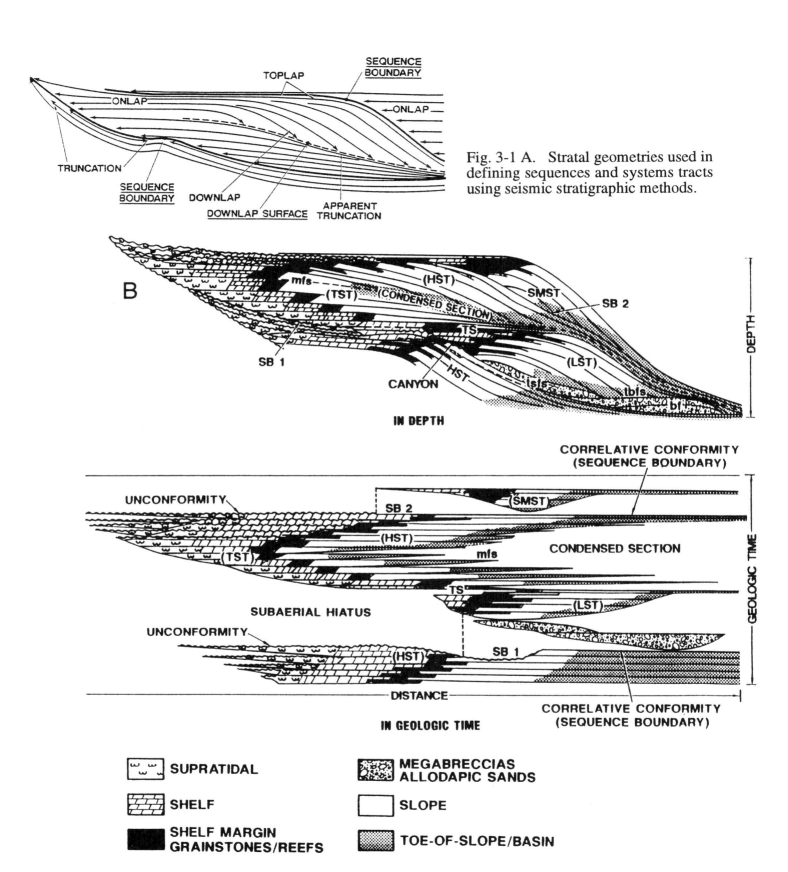

Fig. 3-1 A. Stratal geometries used in defining sequences and systems tracts using seismic stratigraphic methods.

Fig. 3-1 B. Rock-based sequence stratigraphy and facies of a rimmed shelf with a relatively steeply sloping margin plotted in terms of depth (top) and time (bottom). From Vail (1987).

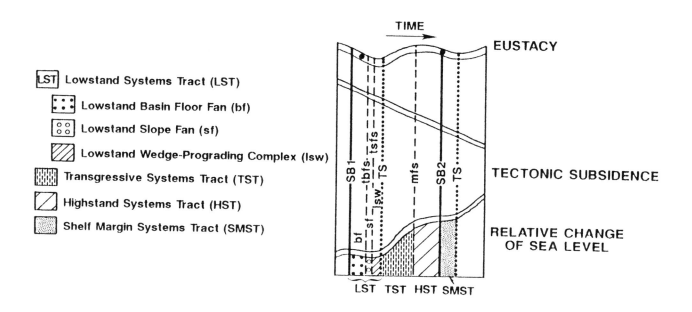

Fig. 3-1 B continued. Interpreted relationship of systems tracts to eustatic sea level curve and relative sea level (After Vail, 1987).

frequency, <u>Milankovitch-driven climate changes</u> and their associated sea level fluctuations. This is indicated by estimates of cycle durations (Heckel, 1980, 1985; Koerschner and Read, 1989; Goldhammer et al., 1990, 1993), by bundling of precessional) carbonate cycles into sets of 5 (eccentricity bundles) (Goldhammer et al., 1990), by spectral analysis of time series constructed from cyclic stratigraphic sections (Bond et al., 1991; Goldhammer et al., 1990, 1993), and by evidence of high frequency sea level drops off the platform forming disconformities, breccias, soils, and vadose fabrics on tops of parasequences (Goldhammer et al., 1990; Koerschner and Read, 1989). Some carbonate cycles (especially some of the thinner ones) are autocyclic and unrelated to eustasy, but instead relate to varying sedimentation rates due to progradation shutting down the carbonate factory, shoaling of islands, or tidal channel switching as proposed by Ginsburg (1971), Goldhammer et al. (1990), Kozar et al. (1990) and Hardie et al. (1991). Note however, that autocycles likely will lack the diagenesis associated with sea level drop off the platform, since the autocycles will shallow to the position of static sea level.

Source rocks, reservoirs and seals are genetically related to their sequence stratigraphic framework. Sequence stratigraphy helps greatly simplify formation-based stratigraphy with its myriad of local names and commonly arbitrary boundaries and groupings of lithologies. From a practical standpoint, the much cited criticism against sequence stratigraphy (that given the limited resolution of biostratigraphy and dating, whether or not depositional sequences and inferred sea level cycles are able to be correlated interbasinally or globally) is of decreased importance to most of us working within an individual basin.

<u>Sequence Boundaries:</u>

Depositional sequences are bounded updip by basal <u>unconformities</u> which show onlap of younger units onto them (Sarg, 1988). In eustatic cycles, the unconformities or sequence boundaries correlate with the time of maximum sea level fall rate which exceeds subsidence rate on that part of the platform.

<u>Type 1 sequence boundaries</u> are characterized by emergence of the platform out beyond the shelf edge and can be caused by eustatic sea level falling faster than the shelf edge is subsiding (Fig. 3-1B). Clearly these boundaries are going to be more common on rimmed shelves than ramps. Note that the type of basal bounding unconformity gives its name to the type of sequence developed (e.g., type 1 sequence).

<u>Type 2 sequence boundaries</u> may have erosional disconformities updip, but sea level does not significantly expose the margin thus sequence boundaries here are conformable; such sequence boundaries may result from eustatic sea level fall being less than subsidence, and such type 2 boundaries are associated with small sea level falls, and are typical of ramps (Figs. 3-1B, 3-2).

In outcrop or core, sequence bounding <u>unconformities</u> (Fig. 3-3) may show:

1. redbeds, green shale, detrital sands and conglomerates overlying carbonates

2. infiltration of sand or green/red shale into fissures, karstic solution cavities and joints in the underlying carbonate; brecciated carbonates adjacent to unconformity and filling caves

3. brown laminated soil-crusts (caliche) and caliche pisolites, sometimes with tepee structures truncated by erosion; updip evaporite solution breccias, and

4. biostratigraphic gap increasing updip

5. on rapidly subsiding platforms or portions of platforms, the sequence boundary may not be a simple unconformity, but can be cryptic, and located within a zone of very thin, cyclic carbonates that show evidence of maximum emergence (e.g. quartz sands/silts in tops or reworked into

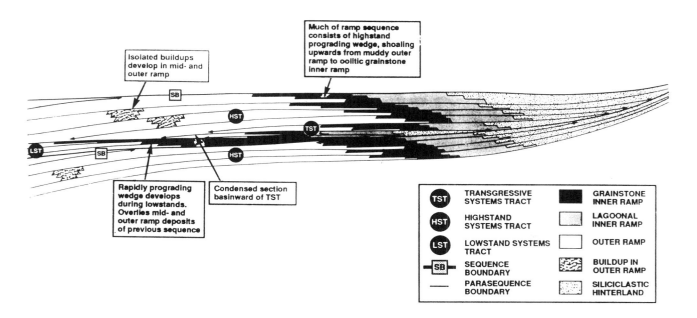

Fig. 3-2. Sequence stratigraphy and systems tracts of a carbonate ramp. Note that there is relatively little thinning downdip and that slopes are relatively gentle. Consequently, systems tracts merely move up or down the ramp rather than being influenced by sea level dropping of the edge of the platform as in rimmed shelves. From Burchette and Wright (1992).

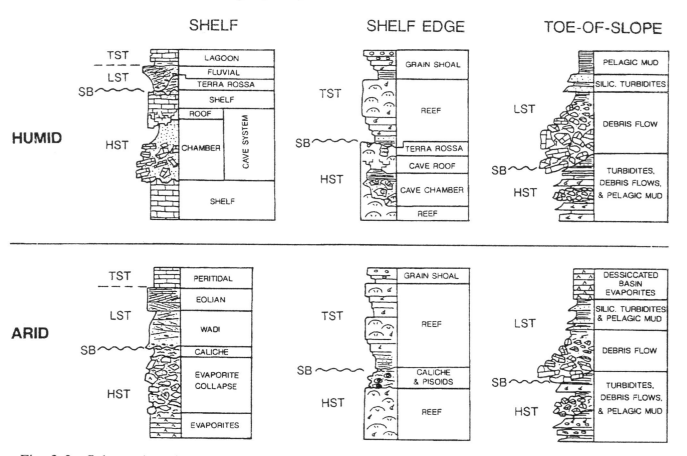

Fig. 3-3. Schematic columns showing lithologies above and below sequence boundaries under arid and humid conditions. From Handford and Loucks (1993).

bases of cycles, clayey paleosols capping the cycles, and tepeed, calichified or brecciated caps on cycles).

6. downdip, the correlative conformity will underlie the low stand deposits (resedimented facies in the case of rimmed margins; in-situ bank/shoal facies on ramp margins).

Note that downdip, sequence boundaries become conformable and pass underneath the low stand deposits of the shelf-margin or ramp-margin. This is because unconformities will form on the shelf during maximum fall which precedes the low stand (Figs. 3-1B, 3-2).

Drowning unconformities (Schlager, 1989; Erlich et al., 1990) may form sequence boundaries on platforms where rate of relative sea-level rise exceeds sedimentation rate and the platform becomes deeply submerged below the euphotic zone, terminating rapid production and accumulation of carbonate by photosynthetic organisms (Kendall and Schlager, 1981) (Fig. 3-4). As noted before, the euphotic zone in the open ocean may extend down to 100 m, but may be as little as 30 m in basins where fine-grained carbonate or clastics are abundant. This drowning surface may be marked by

1. an extremely condensed interval, or

2. sharp hardground surfaces or

3. gradational contacts between shallow and overlying deeper water limestones and shales

4. durations of missing section may be from 1 to 25 million years.

5. downlap of prograding clinoforms of the shelf and slope onto the marine unconformity surface.

Such drowning unconformities can be accompanied by rapid carbonate buildup development along the shelf edge, forming prospective reservoirs.

Low Stand Systems Tract (LST):

These are the first deposits of a depositional sequence, and generally conformably overlie the sequence boundary in the basin and slope. Unconformity-related karsting and soil formation may develop updip on the platform during the low stand.

As Type 1 sequence boundaries are forming as sea level is falling off the shelf edge, a low stand wedge of allochthonous (resedimented) debris from the steeply sloping margin may be deposited, due to failure by oversteepening or slope front erosion or loss of buoyancy of the slope (Fig. 3-1B). On the outer shelf or slope immediately updip from the talus wedge, several stranded parasequences may be deposited at this time related to forced regression (Posamentier et al., 1993).
Failure of the shelf edge may leave a large scalloped embayment, which may be used as an initial indicator of downslope talus deposits. Note that low stand wedges of debris flows and turbidites may be indistinguishable from debris flows resulting from tectonically induced slope failure which would have little relationship to sea level lowstand.

As sea level fall slows and the low stand position of sea level is approached, deposition of a ramp/shelf margin wedge of shallow water carbonate produced in place may occur in the shallowed slope areas, and the slow relative rise in sea level (static sea level plus continued subsidence) will cause onlap of these shallow water facies up the slope and onto the outer platform (Fig. 3-1B). These low-stand buildups may form separate reservoirs from the shelf itself.

Autochthonous low stand ramp margin wedges associated with type 2 boundaries occur commonly on ramps where sea level does not fall below the bank- or shelf margin (Fig. 3-2). This shelf or ramp margin wedge typically consists of several shallow water parasequences localized near the basin margin and commonly resting on deeper water

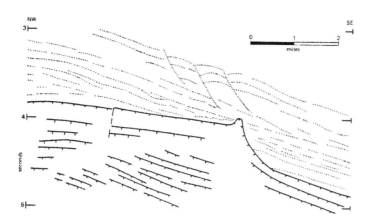

Fig. 3-4. Drowning unconformity of Wilmington platform, off New Jersey, redrawn from Shell Oil Co line. Hachured lines are carbonates, dotted lines are siliciclastics and pelagics. From Schlager, 1989.

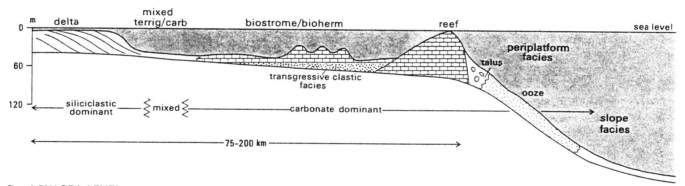

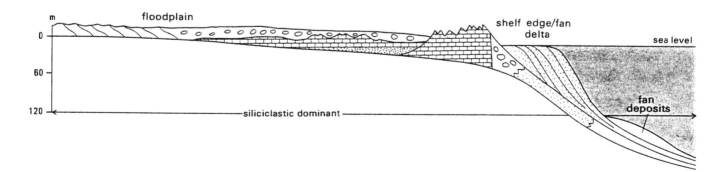

Fig. 3-5. Schematic diagram showing development of high-stand carbonates (A) followed by (B) low-stand karsting and limited siliciclastic deposition on the platform and major deposition of shelf edge and slope/basin fan siliciclastics. Davies, 1989.

facies. The low-stand cycles show only limited onlap onto the platform and limited seaward progradation and downlap onto the basin floor, because during the low stand, sea level is either slowly falling, static or slowly rising, thus background subsidence results in some relative sea level rise. This wedge may be dominated by skeletal bank facies, by ooid shoal facies, or by peritidal cycles.

In arid settings, basins both on the shelf, and downslope from the shelf may become sites of <u>low stand evaporites</u> or evaporitic dolomites. Given sufficient sea water influx, these basins can have high evaporite sedimentation rates of several meters to tens of meters/k.y.

<u>Low-stand siliciclastics</u> may be deposited in the basin as fans as a result of clastic bypass of the exposed shelf (Fig. 3-5). These clastics are transported to the platform edge by sheet-flooding and eolian processes and deposited as fans on the basin floor. They are important in that they form reservoirs and also shallow the basin floor, promoting platform progradation during the next highstand. Larger sea level fluctuations will tend to favor low stand clastic deposition as they will tend to expose all of the shelf or ramp out to the basin margin.

Transgressive Systems Tract (TST):

The TST develops where subsidence or sea-level rise exceeds upbuilding, causing the platform to become incipiently drowned (Schlager, 1981; Kendall and Schlager, 1981). This may progressively move the platform to below the euphotic zone. During drowning, local banks and reefs may keep pace with sea level for a time, forming large pinnacle reefs and platform reefs on the drowning platform (Fig. 3-6).

The TST is separated from the underlying LST along the platform margin by a <u>marine flooding surface (the transgressive surface)</u> across which facies go from shallow water to relatively deeper water facies, and is the first significant marine flooding surface across the outer shelf or ramp (Figs. 3-1B, 3-2). Parasequences of the TST show:

1. onlap onto the sequence boundary (or zone),

2. retrogradational (sedimentation rate less than accommodation rate) or aggradational stacking patterns (sedimentation rate equals accommodation rate). Shallow water facies capping cycles will show progressive backstepping upward in the section, and at the same time there will be backstepping of deep ramp or foreslope facies beneath basin facies.

3. individual parasequences of the TST are the thickest of all the systems tracts and will contain many with very open marine subtidal bases. This is because accommodation rate is high, resulting in maximum water depths during short term sea level rises.

4. TST parasequences have tidal flat facies either absent or restricted to the inner platform, and show the least development of disconformities/soil formation due to the high accommodation rate, which decreases the amount of time cycles remain emergent during short term sea level falls.

The TST is separated from the HST by the <u>maximum flooding surface</u> (MFS). Note that simply picking the base of the deeper water facies directly above shallow water cycles from section to section likely will define a diachronous surface (the individual tops on a series of backstepping, drowning cycles)s and not the actual maximum flooding surface; over much of the outer platform, the MFS may in fact be a fairly cryptic surface/zone within the succession of relatively deep water carbonates. In general, the MFS is characterized by:

1. maximum landward extent of a tongue of deeper water facies extending into shallow water facies

2. hardgrounds; condensed, deep-water, nodular, argillaceous limestone; pelagic carbonate; or chemical

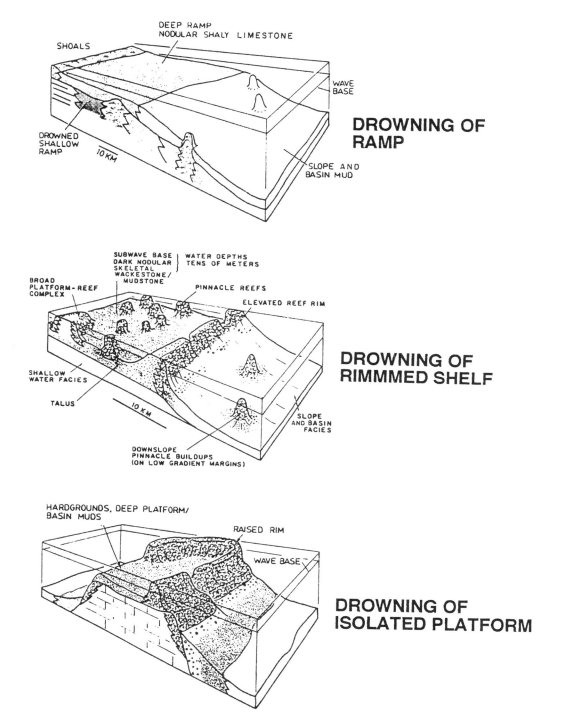

Fig. 3-6. Response of ramps, rimmed shelves and isolated platforms to drowning. Development of raised buildups and overlying deep water seals result in common reservoirs associated with drowning.

sediments (iron, manganese, phosphorite, or sulfide crusts).

3. downlap of parasequences of the HST onto the MFS if there is a significant decrease in sedimentation rate downdip, as with rimmed shelves (Fig. 3-1B). On many ramps, there is little decrease in sedimentation rate with water depth, thus stratal units may show little if any downlapping onto this surface except in extreme downdip positions (Fig. 3-2).

4. on platforms characterized by keep-up, in which the platform top roughly tracks long term relative sea level rise, the MFS may be cryptic, and lies within a zone of relatively thick cycles.

If eustatic, the MFS typically occurs relatively high on the sea level curve, but prior to the actual sea level high stand (Fig. 3-1B). This largely reflects the effects of high rate of sea level rise, coupled with maximum load induced subsidence due to widespread sedimentary loading over the platform due to widespread marine flooding. However, on flat topped isolated platforms, the maximum flooding surface may occur during the time of maximum sea level rise rate.

High-Stand Systems Tract:

Following drowning and onset of decreasing accommodation, deeper water benthonic assemblages are able to build up into the photic zone, assisted by accumulation of deep ramp and slope lime-muds carried in from updip shallow platform areas, along with contribution of fine siliciclastics carried onto the slope. Note that this upward building of the platform following incipient drowning typically is associated with clinoform progradation, and downlap onto the MFS, with the shallower water carbonates building out over their own deeper ramp or foreslope deposits (Figs. 3-1B,3-2). Consequently, very rapid rates of highstand progradation can be achieved if sedimentation rates on the slope are supplemented by fine siliciclastic input either by contour parallel currents, or by high frequency low-stand clastic slope or fan deposition (Meyer, 1989; Sonnenfeld, 1991).

Parasequences of the HST show:

1. marked progradational geometries if sedimentation exceeds accommodation;
if accommodation is high or the platform has high relief, aggradational geometries may be developed during the HST

2. parasequences become dominated upward in the HST by shallower-water and peritidal facies,

3. parasequences become thinner upward (due to progressive decrease in accommodation.

4. parasequences are relatively thin and restricted compared to the TST and LST. In arid settings, progradational HST cycles commonly are dolomitized, capped by evaporites, or may interfinger updip with updip with siliciclastics. In humid settings, HST carbonate cycles may interfinger with prograding nearshore and non-marine clastics.

CHAPTER 4

Some of the material in the following chapters is dealt with in more detail in Read and Horbury (1993).

RESERVOIRS ON PLATFORMS DEVELOPED UNDER SMALL, HIGH FREQUENCY SEA LEVEL OSCILLATIONS (GREENHOUSE CONDITIONS)

Greenhouse cycles may include, the Cambro-Ordovician of the U.S. Appalachians (Koerschner and Read, 1989; Demicco, 1985), Late Silurian and Early Devonian of New York (Goodwin and Anderson, 1985), many Mid to early Late Devonian platforms in Australia and North America (Read, 1973; Wendte and Stoakes, 1982; Elrick, pers. comm., 1992), the Late Permian of the Permian Basin, U.S.A. (Hovorka, Nance and Kerans, 1992; Borer and Harris, 1991) the Mid to Late Triassic (Fischer, 1964; Schwarzacher and Haas, 1986; Goldhammer et al., 1990), the Early Jurassic, Morocco (Crevello, 1991) and the Late Jurassic (Mitchell et al., 1988) and Cretaceous (Strasser, 1988).

Key Features

During greenhouse times there is little or no polar ice, thus:

1. Cycle stratigraphy will be dominated by high frequency carbonate cycles, commonly less than 20 k.y. duration, that formed under low amplitude precessionally driven sea level changes, perhaps with superimposed small 100 k.y. and 400 k.y. sea level fluctuations

2. Carbonate platforms typically are aggraded and <u>flat-topped</u> to gently sloping.

3. Sub-seismic scale peritidal meter scale cycles or parasequences commonly have regionally <u>extensive tidal flat caps</u>

4. <u>High relief buildups</u> such as pinnacle reefs are <u>absent</u> from parasequences on the shallow platform top

5. Parasequences tend to have <u>"layer cake" stacking</u> patterns, arranged into large scale transgressive onlap and regressive offlap patterns within 3rd order depositional sequences

6. Marginal reef/grainstone facies show limited lateral migration from cycle to cycle; may be thick and poorly partitioned, or highly compartmentalized

7. Cycle capping <u>disconformities</u> are relatively poorly developed.

8. Small sea level changes limit vertical migration of ground water tables relative to the platform surface. Arid zone flats generally associated with pervasive early dolomitization of inner platform as brine systems migrate seaward during progradation. Humid settings may develop limited meteoric/mixed lenses beneath exposure surfaces.

Fourth and Fifth Order Sea Level Changes, Greenhouse Times:

Sea level changes during greenhouse times appear to be driven by precession (Fig. 1-5), which today is 19 to 23 k.y. but in the past may have been less than this, and perhaps as little as 15 to 17 k.y. in the Early Paleozoic (Berger et al., 1989). However, it is possible that 10 k.y. and higher frequency cycles also are important. The magnitude of the sea level changes forming precessional greenhouse cycles probably were small, and less than 10 m (Goldhammer et al., 1990; Koerschner and Read, 1989; Wright, 1992) although facies within contemporaneous intrashelf basins suggest somewhat larger sea level changes. Cycles may be bundled into what have been interpreted as eccentricity cycles composed of 5 precessional cycles, which may be observed on Fischer plots as in the Middle Triassic Latemar platform

(Goldhammer et al., 1989). Associated 40, 100 and 400 k.y. sea level changes must be small, otherwise numerous missed beats will punctuate the cyclic succession which will then show relatively poor preservation of the Milankovitch signal, as on large shallow platforms such as the Late Triassic of Hungary (Schwarzacher and Haas, 1986; Balog et al., in press).

Greenhouse Carbonate Cycles:
Cycles typically thin and layer-cake (Fig. 4-1D), extensive sheets with little thickening or thinning. The relative thinness of these meter-scale cycles reflects limited accommodation because sea level fluctuations are small, and accommodation tends to be filled with each precessionally driven sea level fluctuation. This limited accommodation inhibits development of carbonate buildups on the platform top, except during major drowning events associated with pulses of subsidence or long term relative sea level rise. Consequently, any such mounding tends to be localized downslope where there is accommodation space available.

Cycles decrease in abundance updip (and downdip), because flooding of the platform by each sea level beat in part is a function of accommodation rate; the lower the accommodation rate, the fewer the cycles on the inner platform. Downdip, cycles decrease in abundance because higher accommodation here prevents every cycle shallowing to sea level, so that some cycles actually consist of amalgamated subtidal cycles. This difference in the numbers of cycles over the platform makes it difficult to correlate individual cycles with certainty from section to section.

Updip cycles consist of a shallow restricted subtidal facies mosaic overlain by a regressive tidal flat cap (fenestral or microbially laminated carbonate). Less commonly, tidal flat facies form transgressive units above cycle disconformities, especially where ensuing initial sea level rise rates were small. Tidal flat caps and bounding surfaces are inferred to be relatively continuous although this is difficult to prove unless there is continuous lateral exposure. Peritidal facies may disappear close to the ramp or shelf margin where cycles become dominated by grainstone. Widespread tidal flat facies occur on inner parts of ramps and on large, shallow flat-topped platforms because the tidal flats are able to prograde and keep pace with the regressing shoreline as sea level falls. Tidal flat facies do not develop on many small carbonate platforms a few kilometers wide because of the high energy conditions; instead cycle tops can have dolomitic soils developed on subtidal beds (Goldhammer et al., 1990).

Autocycles reflecting local shoaling events may be commonly associated with greenhouse platforms. The small fluctuations of sea level may allow the effects of local shallowing and island formation to generate upward shallowing cycles independent of sea level (Goldhammer et al., 1993).

Cycle Bounding Disconformities:
Cycle capping disconformities are generally poorly developed even though cycle tops may remain emergent for tens of thousands of years, because sea level (and the groundwater table) only falls a short distance below the platform surface. Most of the disconformities are sharp surfaces, with some reworking of indurated cycle caps into the overlying transgressive part of the cycle. Maximum emergence of individual cycle tops occurs during 3rd order sea level fall when regolith, siliciclastic- or caliche tepee-capped cycles can develop (Goldhammer et al., 1990; Koerschner and Read, 1989). Under humid conditions clayey paleosols may develop on tops of emergent cycles.

Depositional Sequences and Stacking Patterns
Sequence boundaries updip marked by multiple exposure surfaces. The best developed exposure surface likely represents the actual sequence boundary although in some cases picking this may be subjective. This sequence boundary actually is a series of merged disconformities from downdip cycles. Sequences generally are conformable downdip, where the sequence boundary is relatively difficult to pick in outcrop or core of highly cyclic sections, so that other techniques need to be used.

Fischer plots (Fischer, 1964) in which cumulative cycle thickness corrected for linear subsidence, is plotted against time, provide a relatively objective way of defining the 3rd order sequences because of the large number of meter scale cycles developed on greenhouse platforms (Read and Goldhammer, 1988; Goldhammer et al., 1993). Actually, Fischer plots are better visualized as plots showing departure from mean cycle thickness vs. cycle number up the stratigraphic section (a proxy that avoids having to estimate time) (Saddler et al., 1993). Thick cycles form during times of increased accommodation rate (e.g. 3rd order rise) whereas thin cycles form during times of decreased accommodation rate (e.g. 3rd order falls). However, examination of stacking patterns using Fischer plots may help define the approximate position of the sequence boundary, which in most cases will be roughly halfway down the falling limb of the Fischer plot, and located within a succession of very thin cycles showing maximum evidence of exposure (quartz rich units, caliches, tepees, regoliths). However, if long term sea level falls off the platform to form a low stand wedge, then Fischer plots on the platform updip from the wedge will show the sequence boundary at the position of the lowest point on the Fischer plot - clearly, you need to know whether there is a low stand wedge to pick the true location of the sequence boundary on the Fischer plots. Note that Fischer plots are best suited to extract accommodation changes from greenhouse cycles because the cycles are numerous and tend to have filled the accommodation during each beat. This is not true for other times in the stratigraphic record.

The TST on greenhouse platforms will be made up of a succession of relatively thick cycles with well developed subtidal units that are the most open marine of all the systems tracts. Cycles may lack laminite caps, but instead shallow up to a planar to scalloped intertidal to subaerial erosional surface, especially on the outer platform. Commonly the TST will have the least early dolomite of all the systems tracts. Evidence of exposure will be minimal compared to other systems tracts.

The HST consists of a succession of intermediate thickness cycles, with well developed tidal flat caps on more restricted subtidal units. HST cycles commonly are completely dolomitized on arid platforms. Disconformity-capped cycles are common.

The LST cycles may be of similar thickness to those of the HST, but will generally occupy only the seaward margin of sloping platforms with an updip sequence-bounding unconformity. On flat topped platforms with high accommodation, the LST may extend across the platform, and show similar cycles to those of the HST. LST cycles on land-attached platforms may show increased siliciclastic input compared to other systems tracts. The siliciclastics may be transported across the platform by wind and sheet-flooding, but generally become reworked to form thin transgressive marine sand sheets at the base of each cycle.

Coarse Grained Margin Facies:
Skeletal and oolitic grainy facies are restricted to the high energy margins of greenhouse ramps and rimmed shelves. Grainy units may only be a few kilometers wide, and show limited updip or downdip migration from cycle to cycle because high frequency sea level changes are small. However, these facies may show large scale migration associated with backstepping or prograding of systems tracts of depositional sequences. Grainy platform margin facies lack subaerial breaks, and may show little vertical partitioning, where high frequency sea level changes are small. Higher amplitude sea level changes will tend to cause platform margin grainstones to be partitioned vertically by cycle-capping, tighter, muddy lagoonal and peritidal facies. On seaward parts of ramps, grainy units may be partitioned by muddy deep ramp carbonates/shaly carbonates at bases of cycles.

Margins generally do not show evidence of clinoform bundling of shelf edge strata in which facies demonstrate considerable vertical shifts - where they do it would imply the presence of reasonably high 4th order sea level changes superimposed on the higher frequency signals.

Reservoirs and Diagenesis:
Arid zone cycles have restricted, commonly oolitic and cryptalgal mound facies and intertidal laminite caps; cycles are partly to completely dolomitized by brines sourced from the tidal/supratidal surface (Fig. 4-4).

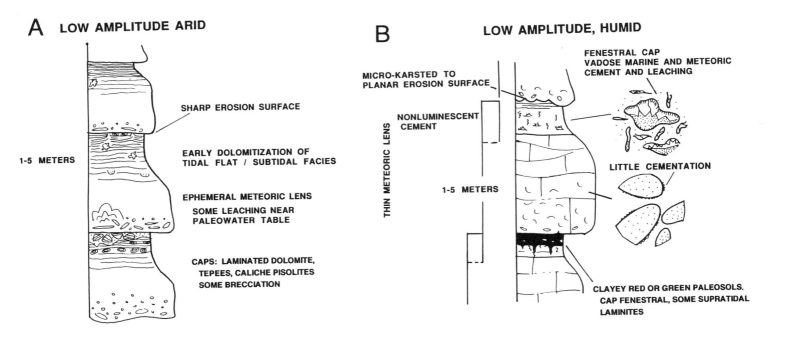

Fig. 4-1 A. Idealized stratigraphic columns showing carbonate cycles generated under arid climate and low amplitude sea level fluctuations. Initially, only the tidal flat laminites are dolomitized, but with repeated progradation, cycles become completely dolomitized and there may be local leaching of aragonite beneath the flats.
B. Idealized stratigraphic column of carbonate cycles generated during low amplitude sea level fluctuations in a humid climate. Marine vadose, meteoric vadose and possibly shallow phreatic cements are typically confined to the fenestral tidal flat cap which may show some micro-karsting. Laminites relatively scarce.

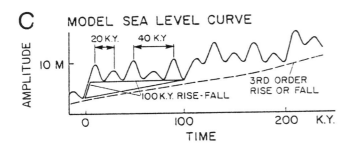

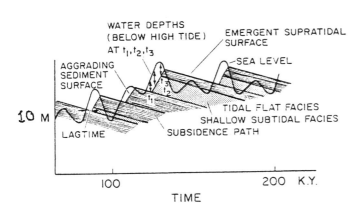

Fig. 4-1 C. Computer model illustrating how carbonate cycles may form under small Milankovitch oscillations in sea level. Top: sea level curve and bottom, track of sediment surface and facies developed plotted on a time vs distance/stratigraphic thickness plot (Koerschner and Read, 1989).

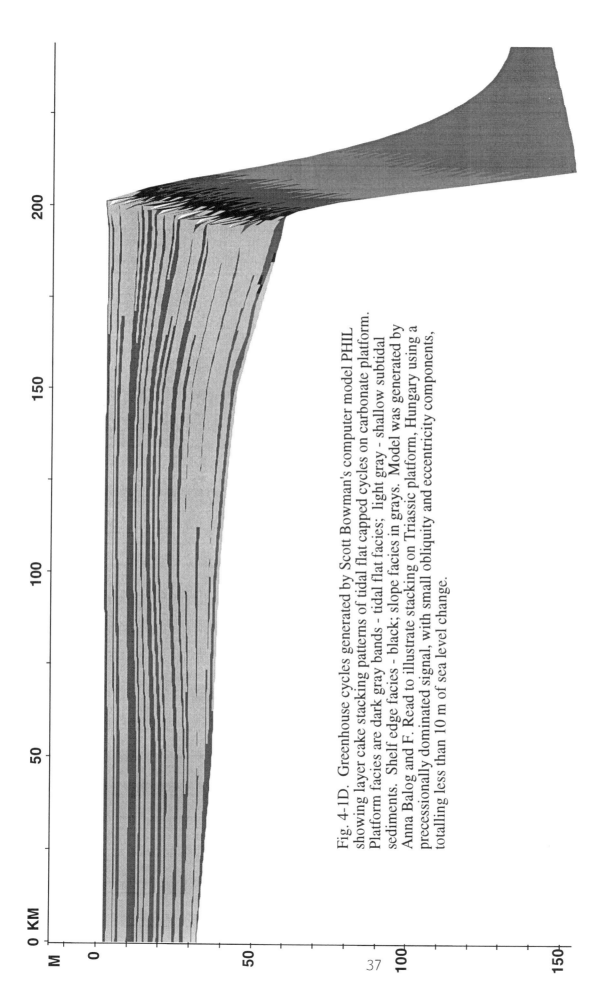

Fig. 4-1D. Greenhouse cycles generated by Scott Bowman's computer model PHIL showing layer cake stacking patterns of tidal flat capped cycles on carbonate platform. Platform facies are dark gray bands - tidal flat facies; light gray - shallow subtidal sediments. Shelf edge facies - black; slope facies in grays. Model was generated by Anna Balog and F. Read to illustrate stacking on Triassic platform, Hungary using a precessionally dominated signal, with small obliquity and eccentricity components, totalling less than 10 m of sea level change.

Reservoir facies can occur in fine laminated dolomite caps where intercrystal porosity is preserved, in cycle capping siliciclastics, or in interparticle and intercrystal/intergranular porosity in variably dolomitized, subtidal grainstone/packstone shoal facies; plugging of grainstone porosity by sulfate minerals can leave only the fine dolomites permeable. Reservoir thickness is rarely more than 2 to 3 m, and sabkha evaporites can provide extensive seals. This results in strongly stratified, layered reservoirs. The cycles are evident even in highly dolomitized core, and wireline logs are useful in identifying and correlating the cycles by picking out the evaporite cap or feldspar rich silty-dolomite caps.

Humid zone cycles commonly have fossiliferous subtidal facies, fenestral and poorly laminated intertidal facies, and rare supratidal laminite caps (Fig. 4-1B). Tops are planar to microkarsted, and can have associated shale. Aragonite fossils commonly are leached and fenestral porosity can be occluded by fibrous marine and vadose/phreatic sparry calcite cement. These cycles will be most porous in their lower part but cycle tops will have more primary porosity than their arid counterparts, such that internal seals will be poorly developed. Reservoirs will be layered and have permeability baffles. More homogeneous and better reservoirs will occur downdip in subtidal grainstone complexes. Cycles are subseismic scale and should be identifiable by modern wireline logs particularly if cycle boundaries have a clastic component.

Example 1.-Middle Cambrian-Early Ordovician Cycles, Eastern U.S.A.:

Cambro-Ordovician cycles in North America host major petroleum and lead-zinc deposits whose localization is controlled mainly by the overlying unconformity. These extensive tidal-flat capped, sheet-like meter-scale cycles of the Middle Cambrian-Early Ordovician passive margin (Elbrook to Knox-Beekmantown Group) (Fig. 4-1 to 3) probably were formed under low amplitude Milankovitch precessional and eccentricity driven sea level oscillations (Fig. 4-1C), evidenced by spectral analysis, and which perhaps were modified by autocyclic processes (Koerschner and Read, 1989; Osleger and Read, 1991; Bond et al., 1991; Montanez and Read, 1992a,b; Goldhammer et al., 1993; Montanez and Osleger, 1993). A possible sub-Milankovitch signal of about 10 k.y. also is evident (Bond et al., 1991).

Cycles are 1 to 5 m thick (Fig. 4-3). Platform interior cycles consist of subtidal muds and rare small SH-LLH stromatolites overlain by laminite. Further seaward, cycles contain thin basal oolite and intraclastic beds, with microbial mounds (thrombolites and in more open settings, digitate 'algal' mounds) and regionally extensive caps of laminated dolomite with local, silicified evaporite nodules reflecting semi-arid climate (Mazullo and Friedman, 1975; Friedman, 1980). In the most seaward settings, cycles lack laminites and instead contain muddy subtidal carbonates with storm deposits, overlain by ooid grainstone or bioherms, some of which have erosional tops. Within intrashelf basins on the Mid- to Late Cambrian shelf, meter-scale cycles are of deeper marine shale up into small scale hummocky quartz-pellet silts, capped by skeletal/radial ooid grainstone and flat pebble conglomerate (Markello and Read, 1982).

Third order sequences are not easily defined on the basis of sequence-bounding unconformities in the Cambro-Ordovician, because the sequences are mainly conformable over much of the shelf. Fischer plots (Fig. 4-2B and 4-3) define several 3rd order sequences that are traceable interbasinally, suggesting that they are defining eustatic 3rd order cycles (Read and Goldhammer, 1988; Osleger and Read, 1991; Goldhammer et al., 1993; Montanez and Osleger, 1993). Third order sequences (Fig. 4-3) are evident on the outer platform by thick, shallowing-upward, limestone-rich TST cycles overlain by restricted, thin dolomite cycles of the HST and LST, some of which contain quartz sand and shale in the caps or reworked sands in bases of overlying cycles; Some of the late HST and LST cycles have brecciated tops (Hardie, 1986; Read and Goldhammer, 1988; Koerschner and Read, 1989) (Figs. 4-2,3). Zones of cycles showing intense subaerial exposure define the sequence boundary (Montanez and Osleger, 1993). In the subsurface Rome

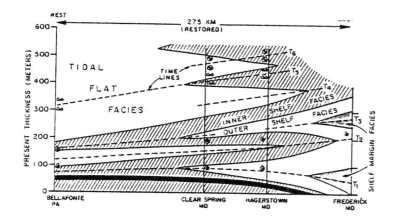

Fig. 4-2 A. Regional stratigraphic cross-section of the Early Ordovician Beekmantown Group, Appalachians, showing depositional sequences in Pennsylvania and Maryland (Hardie, 1989). The transgressive-regressive cycles correspond to the portions of the Fischer plot below labelled O-3 to O-6.

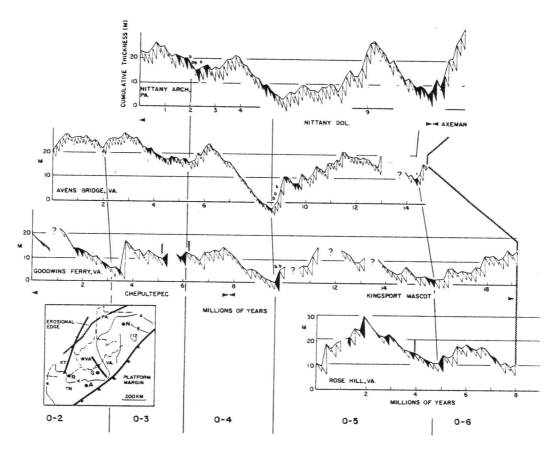

Fig. 4-2 B. Fischer plots which graph changes in accommodation space for the Early Ordovician rocks, Pennsylvania to Virginia. Inset shows location of measured sections - N is Nittany Arch, PA., A is Avens Bridge, VA., G is Goodwins Ferry, VA and R is Rose Hill, VA. On the Fischer plots, small vertical lines are cycle thicknesses, and short inclined lines sloping down to right are linear subsidence paths for each cycle. Increases in accommodation are shown by wavy line sloping up to the right, whereas decreases in accommodation are shown by lines sloping down to right. Note the relatively good correlation between the plots, and the tendency for quartz-rich cycles (black) to occur on the long term falls. The plots probably reflect eustatically driven accommodation changes, because they can be correlated over very large regions. From Read and Goldhammer, 1988).

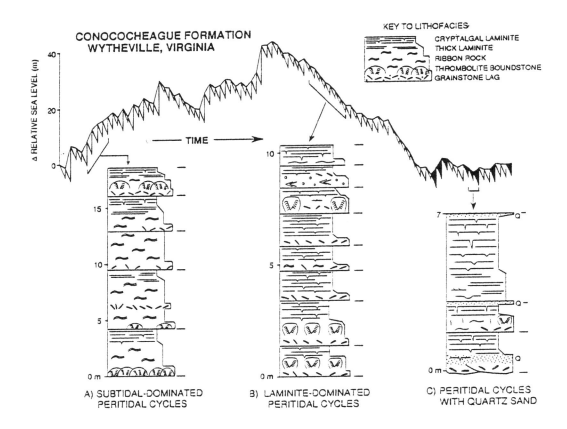

Fig. 4-3. Fischer plot of Late Cambrian cycles in the Virginia Appalachians, Wytheville, measured by Koerschner and Read (1989), from Osleger and Read (1991). Plot shows well defined 3rd order rise-fall, with quartz sandy cycles on the low stand. Stacking patterns of representative cycles are shown below their position on the plot. Note the difference in scales on the 3 columns, and the thick open marine cycles that developed on the rise, vs the thin peritidal cycles that formed on the fall.

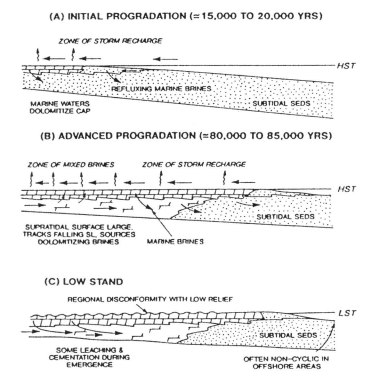

Fig. 4-4. Schematic diagram of the hydrology of an arid tidal flat, modified from McKenzie, Hsu and Schneider (1980) and Montanez and Read (1992b). The carbonate cycle consists of a lower, subtidal unit and a capping unit of tidal flat laminite. Undolomitized limestone shown by stippled pattern; dolomite is shown by standard dolomite "brick" pattern. HST and LST are highstand sea level and lowstand sea level. A) Highstand situation during initial progradation when dolomitizing brines are generated by evaporation of tidal and storm flooding of flats. B) Situation during falling sea level after considerable progradation of flats. Updip, the supratidal surface becomes a disconformity, and continental waters may mix with marine brines generated by storm recharge. Much of the inner platform section becomes dolomitized by refluxing brines. C) At the low stand of sea level, much of the supratidal surface becomes a regional disconformity, and minor leaching of subtidal facies beneath the laminite cap may occur.

Trough, some low stand systems tracts have evaporite-rich dolomite cycles (Ryder et al., 1992).

Most of the dolomitization in the Knox Group was early and synsedimentary, which is indicated by dolomite clasts locally reworked across meter scale cycle boundaries into overlying subtidal limestones and formed under semi-arid climate (Mazullo and Friedman, 1975; Montanez and Read, 1992a). The dolomitization is virtually complete on the inner platform, affecting both tidal flat caps, and subtidal facies. On the outer platform, only HST/LST are highly dolomitized; dolomite in the TST is restricted to laminite caps of cycles. Undolomitized subtidal parts of some cycles preserve minor evidence of meteoric diagenesis (aragonitic cements, meniscus cement, minor crystal silt, and some leaching of ooids). The laminated caps of cycles were dolomitized by brines beneath extensive tidal-supratidal flats (Fig. 4-4). The dolomites initially appeared to have had heavy oxygen isotope values and low Fe and Mn and moderate intercrystal porosity. These dolomites were subsequently reset to varying degrees by replacement and overgrowth by burial dolomite resulting in non porous dolomites that geochemically resemble burial dolomites (Montanez and Read, 1992b).

Example 2. - Late Permian San Andres-Grayburg Cycles, Permian Basin, U.S.A.:

Late Permian (Lower Guadalupian) San Andres and overlying Grayburg dolomites are major oil reservoirs in the Permian Basin of West Texas (Major, Bebout and Lucia, 1988). The San Andres and Grayburg formations (Longacre, 1980; Harris and Stoudt, 1988; Chuber and Pusey, 1985; Hovorka et al., 1993) contain several depositional sequences (Kerans et al., 1992)(Figs. 4-5 and 4-6). Lower parts of depositional sequences are non-cyclic (Fig. 4-6A), open marine carbonates, or poorly cyclic, muddy carbonates with grainstone caps. On the inner ramp, these are overlain by thick successions of cyclic, 1 to 10 m thick muddy carbonate cycles with fenestral caps, and are thoroughly dolomitized. Toward the ramp margin, cycles are dolomitized packstone to grainstone (oolitic, peloidal or skeletal), some of which coarsen upwards and developed low relief over the crestal shoal, and are capped by erosion surfaces or local tidal flat fenestral facies (Figs. 4-6B,C). The upper parts of depositional sequences contain early dolomitized cyclic peritidal facies which form the superposed and marginal seals for the reservoirs. These peritidal facies are fenestral laminated, pisolitic and commonly have tepee structures, interbedded anhydrite, and may be capped by regional eolian siliciclastics that commonly are reworked during cycle transgressions. The parasequences were formed by small, high frequency fluctuations in sea level in relatively arid settings. Cycle facies show well defined backstepping relations in the TST and offlapping relations in the HST (Figs. 4-7A,B).

Near the shelf edge, prograding highstand San Andres facies show compartmentalization of shelf edge facies, related to 4th order eustasy (Sonnenfeld, 1991). The San Andres prograding margin shows a series of imbricate 4th order wedges of grainstone grading down-slope into wackestone and bafflestone. Grainy facies are not interconnected updip but are compartmentalized. Although most of the platform cycles formed under low amplitude sea-level changes, shelf-edge facies reflect mainly 4th to 5th order fluctuations of 10 to 20 m based on the downward shift in shelf edge facies within each 4th order cycle (Fig. 4-8). Rapid progradation was aided by 4th order low-stand deposition of siliciclastics over which 4th order highstand facies prograded.

The San Andres-Grayburg Formations underwent early dolomitization probably from hypersaline brines from tidal flats (Major et al., 1988). The high frequency cyclicity has generated porous reservoir zones up to 15 m thick, separated by relatively impermeable seals. Some reservoirs occur in dolomitized grainstone with interparticle and intercrystal porosity (Major et al., 1988; Hovorka et al., 1993). These grainstones appear to have undergone diagenesis in meteoric lenses beneath local tidal flats or bar crests; these meteoric lenses probably were related to high frequency sea level fluctuations that formed the

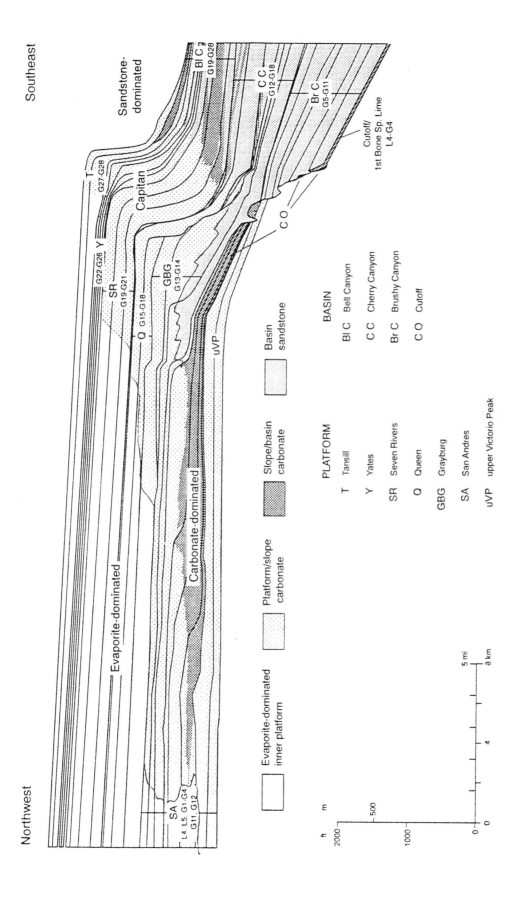

Fig. 4-5. Part of the Permian sequence stratigraphy of the Delaware Basin, Texas and New Mexico. L=Leonardian, G=Guadalupian. Broad ramps with decoupled basinal siliciclastics typify sequences L4-G-12. G-5 to G11 are low-stand clastics most of which bypassed the shelf. Contraction of carbonate facies tracts, increased preservation of siliclastics on the platform and coupled siliciclastic-carbonate facies tracts with intermediate and high order cycles characterize sequences G13-G-28. Note the upward transition from ramps to a rimmed shelf with time. Much of the rim appears to have been relatively deep, perhaps reflecting saline shelf waters. From Kerans et al., 1992.

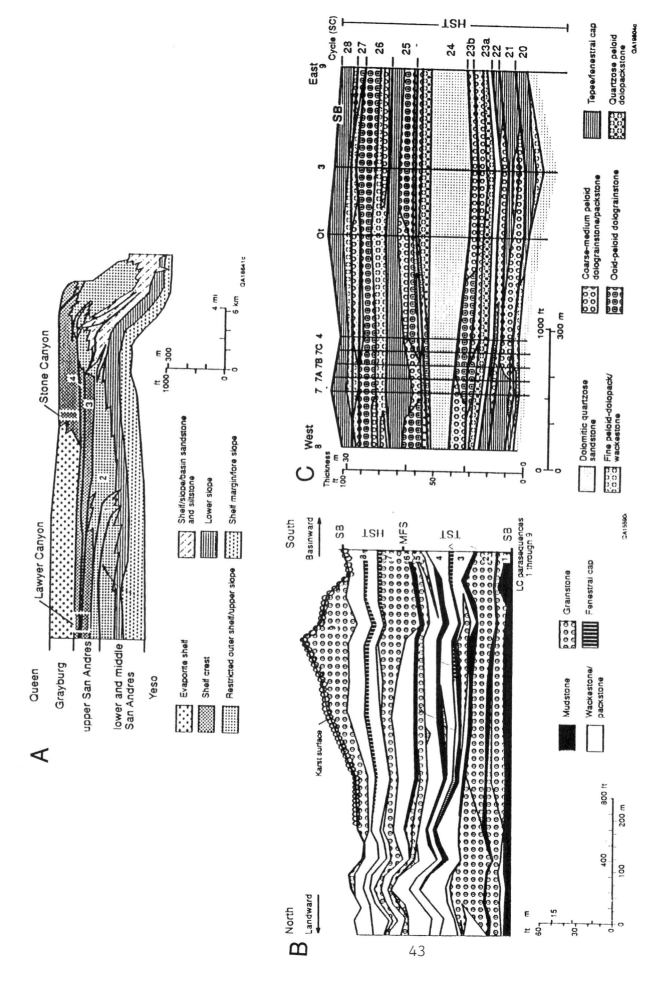

Figure 4-6. Low amplitude Late Permian cycles of the San Andres-Grayburg formations, Guadalupe Mountains, New Mexico (from Hovorka, Nance and Kerans, 1993). A. Regional cross section. B. Detailed cross section of basal upper San Andres depositional sequence, Lawyer Canyon, showing mudstone to wackestone/packstone to grainstone to fenestral capped cycles, and regional karstic surface at the sequence boundary. C. Detailed cross section of Grayburg Formation showing tepee/fenestral or dolomitic sandstone capped grainstone cycles of the HST and the sequence boundary. Note that in both the cross sections, diagenesis associated with paleo-groundwaters during high frequency sea level drops modified and helped to preserve the porosity and in some cases caused enhanced permeability.

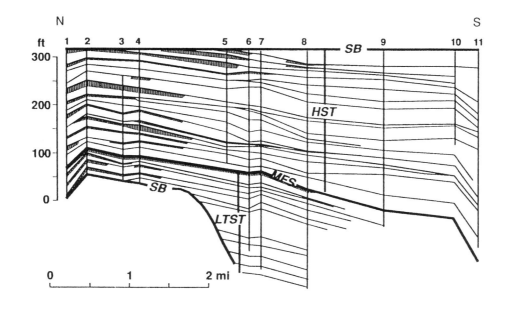

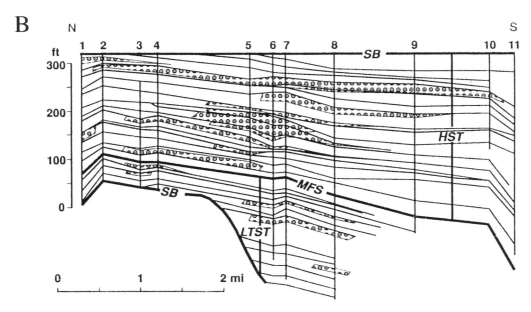

From Kerans and Nance, 1991

Fig. 4-7. Stacking of selected cycle facies within depositional sequences near the ramp margin of the Guadalupian Grayburg Formation, Delaware Basin. A. Distribution of tidal flat fenestral and tepee facies (vertical shading) within Grayburg cycles; cycle boundaries shown by thin lines. B. Distribution of ooid grainstone facies within Grayburg cycle framework. Note the backstepping to forestepping relations of these facies within the sequence, relative to the margin. Note that the LST actually may contain a lower sequence (Kerans, pers. comm., 1994).

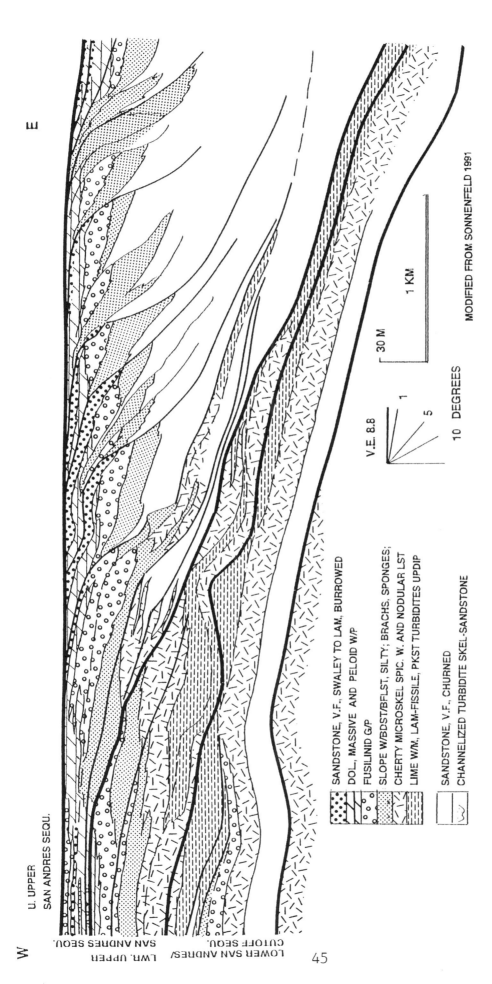

Fig. 4-8. Cross-section of San Andres Formation, Last Chance Canyon, Delaware Basin, grossly simplified from the detailed study of Sonnenfeld, 1991. Sequence boundaries separating the 3 San Andres sequences marked with heavy lines. Note the strongly imbricated, ramp margin units of the prograding HST of the upper sequence, and compartmentalized potential reservoirs. These appear to reflect 4th order sea level fluctuations of perhaps 10 to 20 m.

parasequences (Hovorka et al., 1993). This caused leaching of the metastable grains, mineralogic stabilization, and partial intergranular cementation which helped preserve porosity against compaction. However, many grainstones became tight due to plugging by sulfate cement (Longacre, 1980; Chuber and Pusey, 1984). Many reservoirs also commonly occur in muddy skeletal carbonates that were dolomitized and leached; porosity in these is typically in fine grained dolomite with skel-moldic and intercrystalline porosity (Longacre, 1980; Harris and Stoudt, 1988).

Example 3.- Late Permian Yates Carbonate-Siliciclastic Cycles and Arid Zone Diagenesis, Permian Basin, U.S.A.:

Regressive siliciclastic units capping peritidal cycles in the Late Permian (Late Guadalupian) Yates Formation, Permian Basin, form significant hydrocarbon reservoirs (Borer and Harris, 1991) (Figs. 4-5,9). Cycles are well defined on the basis of gamma-ray logs and are a few meters to 20 m thick (Fig. 4-10). On the inner shelf, the cycles are dominated by anhydrite, halite and red argillaceous siltstones. These pass into middle and outer shelf dolomite-siltstone-sandstone cycles. The carbonate parts of cycles are lagoon/tidal flat facies of the middle shelf- to pisolite-shoal facies on the outer shelf.

The most porous facies are in the clastic dominated mid-shelf. Evaporite-dominated inner shelf and carbonate-dominated outer shelf facies typically are nonporous.

The carbonate-siliciclastic cycles are interpreted to have formed under 4th order (100 to 400 k.y.) low amplitude eustasy. Fischer plots show a bundling of cycles into groups of 4 suggesting 100 k.y. cycles bundled into 400 k.y. cycles (Fig. 4-10C). The sands tend to split downdip into higher frequency units. The Permian carbonates were pervasively dolomitized by downward and seaward migration of brines generated from the evaporitic shelf.

Example 4. - Middle to Late Devonian Humid Zone Carbonates, Canada, Swan Hills - Judy Creek:

The Swan Hills-Judy Creek platform reef complex in Western Canada is 67 m thick, 122 sq. km in area and had original oil in place of 830 million barrels (Wendte and Stoakes, 1982). The complex consists of stromatoporoid reefal rim and foreslope, and platform interior lagoonal facies and formed during drowning of the seaward part of the underlying regional platform (Fig. 4-11). It contains 4-5 regionally traceable megacycles 10 to 18 m thick, each characterized by regional tidal flat caps, and backstepping of the margin (Wendte, 1992) (Fig. 4-12).

The lower 4 megacycles have smaller (1 to 5 m thick) cycles, most of which are traceable across the buildup, but whose tidal flat caps are less continuous. The small cycles consist of dark, muddy Amphipora rudstone overlain by lighter colored, mud-poor Amphipora rudstone with packstone/grainstone matrix, and cycle caps of fenestral and crinkly laminated pellet mudstone/packstone and beach lime sand/gravel facies and thin green shales. Although the tidal flat facies are not traceable between all wells, the deepening events marking most cycle boundaries are traceable. The megacycles associated with backstepping may be 4th order cycles (100 to 400 k.y.) while the smaller more discontinuous cycles within the megacycles may reflect low amplitude precessional/obliquity forced sea level changes.

The cycles were modified in a non-evaporitic setting, indicated by lack of early dolomite and evaporites. Fresh water vadose diagenesis of the lagoonal facies has produced exposure zones 30 cm to 2 m thick along with secondary vug formation, and moldic and channel porosity that may be lined with pendant fibrous calcites; the associated green shales may be paleosols, or storm transported siliciclastic mud (Walls and Burrowes, 1985; see also Read, 1973 for similar diagenetic effects in the correlative Pillara cycles, Canning Basin, Western Australia). These diagenetic events likely were related to fresh water diagenesis during high frequency sea level falls that exposed the platform top. Porosity in the Judy Creek reef

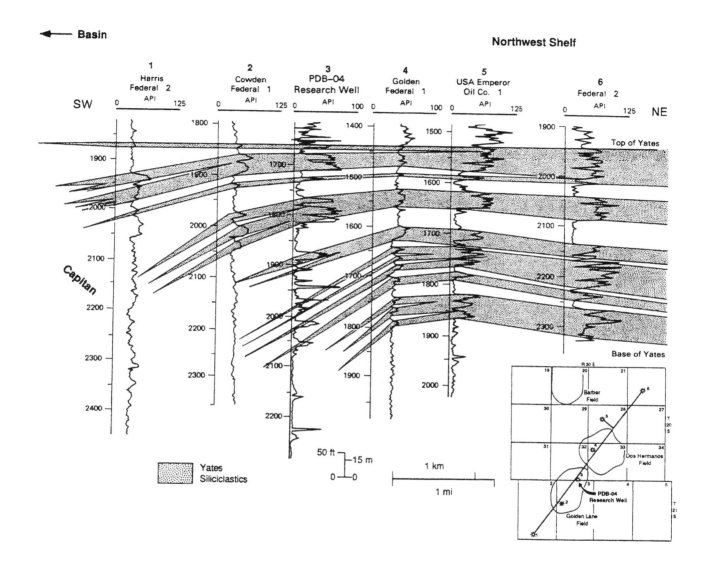

Fig. 4-9. Dip-oriented gamma-ray cross section of the Late Permian Yates Formation in the subsurface of the Northwest Shelf. On the platform top, the beds are relatively flat-lying but they have a gentle slope towards the margin. The interbedded siliciclastic-carbonate cycles show a large scale bundling suggestive of 4th order eustasy. Borer and Harris, 1991.

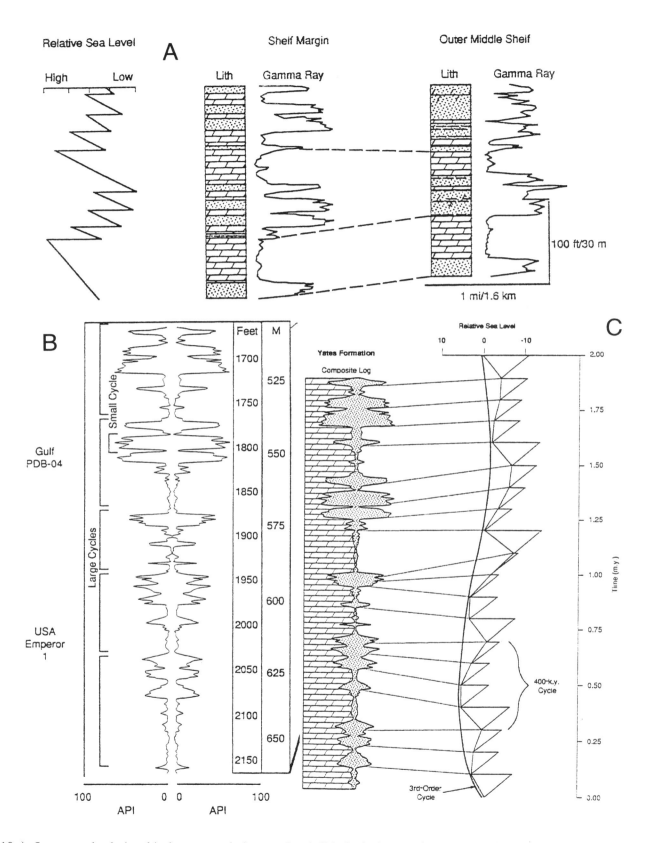

Fig. 4-10 A. Interpreted relationship between relative sea level, lithologic logs and gamma-ray logs, Yates Formation, Northwest Shelf. B. Composite log showing gamma-ray signature of the Yates cycles; the gamma-ray signal is plotted against its mirror image to emphasize bundling of 4 small cycles into large cycles shown. C. Fischer plot constructed from gamma-ray log, showing bundling of 4 small cycles into larger cycles, suggesting 100 k.y. cycles bundled into 400 k.y. cycles. All from Borer and Harris, 1989.

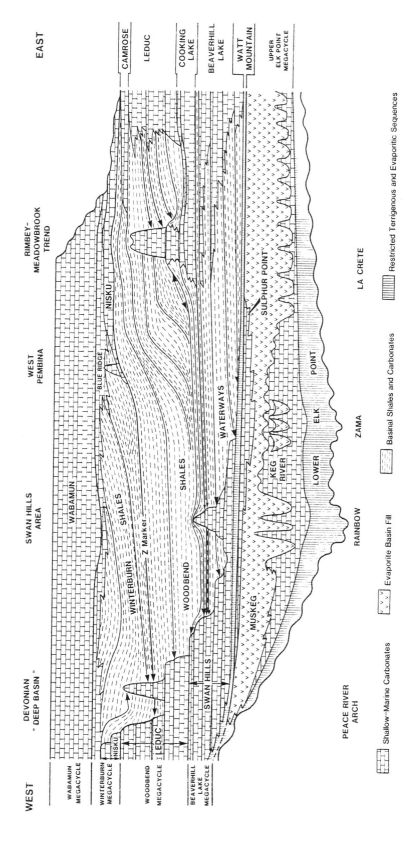

Fig. 4-11. Schematic cross-section illustrating Devonian megacycles or major sequences, along with distribution of major facies. From Wendte, 1992.

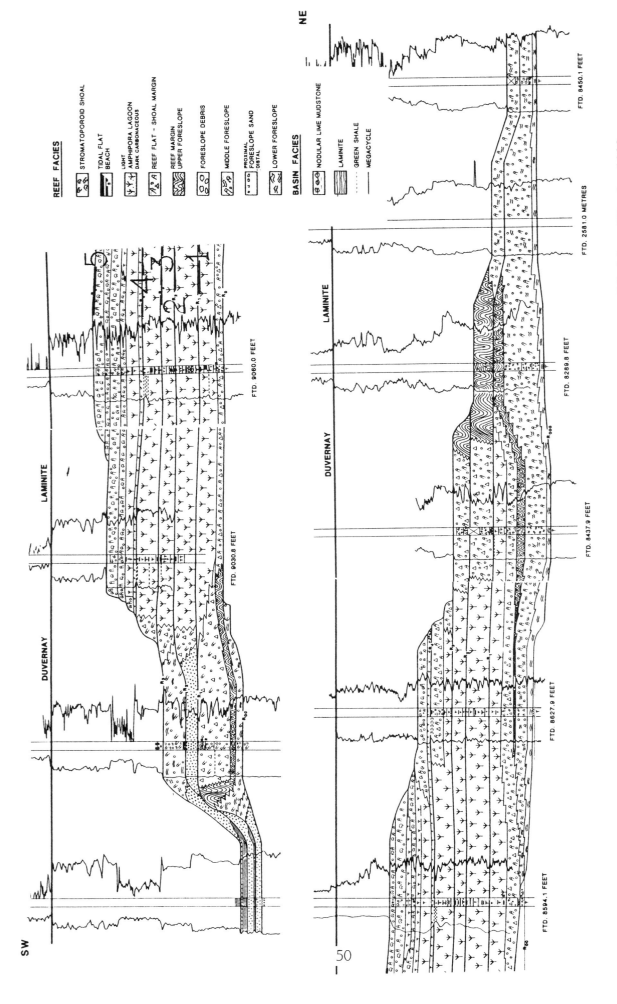

Fig. 4-12. Sequence-facies cross-section across Judy Creek reef complex, western Canada. Modified from Wendte, 1992.

complex is mainly primary and is most abundant in the grainy or reefal margin facies. Porosity of the muddy lagoonal facies tended to be occluded during stylolitization and burial cementation, but discontinuous porosity is associated with shoaling grainy facies of lagoonal cycles.

Example 5.- Tethyan Triassic Cycles: (Hungarian cycles - A. Balog)

High frequency, meter-scale carbonate cycles dominate platforms in the Tethyan Triassic of the Alps (Goldhammer et al., 1990; Fischer, 1964) and Hungary (Schwarzacher and Haas, 1987). Cycle durations are believed to be 20 k.y. based on overall age constraints, but more recently it has been suggested on the basis zircon dating that these cycles may be much shorter duration and sub-Milankovitch (Mundil et al., in press).

Middle Triassic cycles of the small Latemar platform consist of subtidal skeletal lithoclast grainstone overlain by dolomitized grainstone with much evidence of vadose diagenesis and calichification (Fig. 4-13). LST and HST cycles on the platform are arranged into bundles of 5 cycles for each megacycle which are clearly evident on Fischer plots and are presumed to record a relatively complete record of sea level fluctuations (Fig. 4-14). TST cycles are thick and dominated by subtidal packstone grainstone with much marine cementation, and only minor evidence of subaerial exposure about every 10 m. Missing beats due to incomplete shallowing are common (Fig. 4-14). Very thin, tepee-disrupted cycles, typically capping bundles (megacycles) of 3 to 4 cycles occur during minimum accommodation (falling limb of the 3rd order relative sea level curve) (Fig. 4-14). The tepees are characterized by intense vadose diagenesis and expansive cementation. A few thick tepee zones involve more than one megacycle. The tepee-capped megacycles result from one or more missing beats per megacycle, the missing beats occurring on the falling limb of the 100 k.y. (?) cycle. The lack of tidal flat caps on cycles of the small Latemar platform probably is due to wave/tidal current sweeping of the platform top as it shallows each precessional cycle. This incomplete filling of accommodation during each precessional beat thus allowed complete bundles of 5 cycles per megacycle during times of favorable accommodation (LST and HST).

In contrast to the small Latemar platform cycles, the large Late Triassic platform in Hungary and the Italian Alps contains abundant tidal flat facies in the cyclic stratigraphy (Fig. 4-15) (Goldhammer et al., 1990; Schwarzacher and Haas, 1987; Balog et al., in press). Fischer (1964) initially documented these classic Lofer cycles as consisting of transgressive laminites overlain by subtidal limestone capped by a disconformity and paleosol. However, typical cycles also include classic regressive cycles (subtidal carbonate-regressive laminite-disconformity/paleosol) and symmetrical cycles (transgressive laminite-subtidal carbonate-regressive laminite-disconformity/paleosol) (Fig. 4-15) (Goldhammer et al., 1990; Schwarzacher and Haas, 1986). The abundant laminites in the cyclic stratigraphy is due to the large area and low energy of the platform; this favored accumulation of fine sediment in broad tidal flats that built out across the subtidal platform during each shallowing event. Cycles only rarely show characteristic 5 to 1 bundling, rather the bundles only contain 3 or 4 inferred precessional cycles/eccentricity cycle (Balog et al., in press). Some sections also show a weak large scale bundling of 4 small scale bundles into larger scale bundles of 12 to 15 cycles; these poorly defined larger bundles may be 400 k.y. bundles. A similar Milankovitch signal is evident in the offshelf facies of the Dachstein Limestone (Reijmer et al., 1993). Several large depositional sequences (1 to 5 m.y. duration) are evident on Fischer plots. Numerous missing beats in the platform stratigraphy are indicated by abundant paleosols throughout the sections. Paleosol development ranges from thin, incipient types formed during 20 k.y. sea level falls, to thick composite forms which appear to mark numerous missed beats (Fig. 4-15). The abundant missed beats in the platform stratigraphy results from the aggraded character of the Late Triassic platform, on which all the available accommodation was filled during each precessional highstand. Because there was no accommodation unfilled from the previous sea level beat, sufficient new accommodation had to be created by subsidence and sea level rise for the next sea level beat to flood the platform and make a cycle.

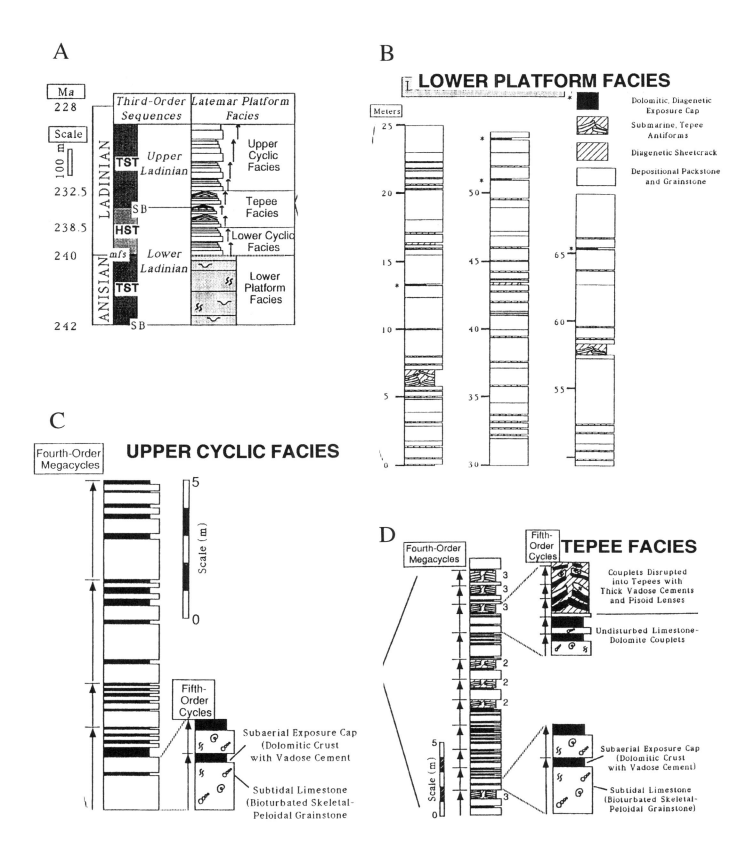

Fig. 4-13. A. Generalized stratigraphic column of the Latemar Platform. B. Dominantly subtidal cycles of the Lower Platform. C. Highly cyclic units from the Upper Cyclic Facies. D. Cycles from the Tepee Facies. Slightly modified from Goldhammer et al., 1993.

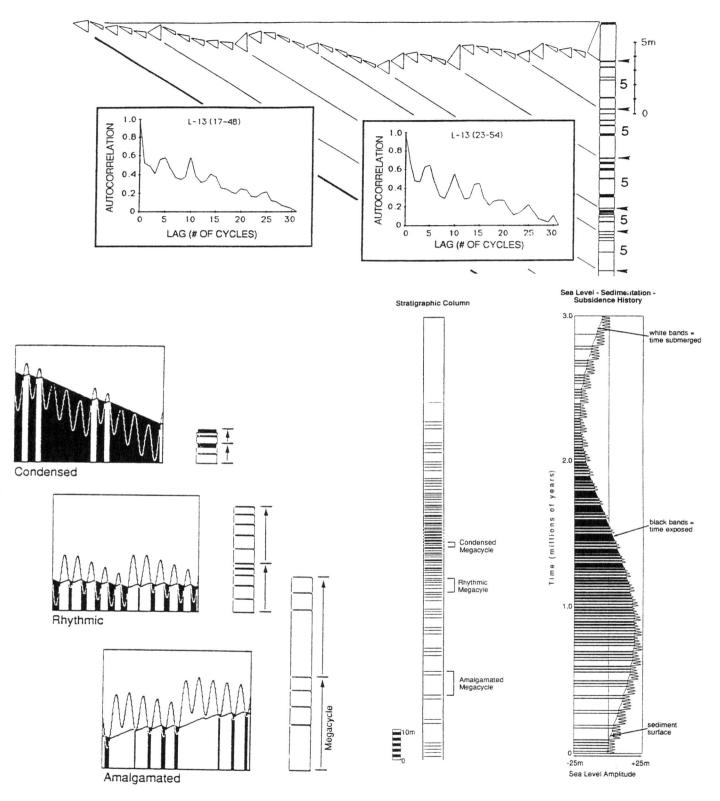

Fig. 4-14. Top: Fischer plot showing 5 to 1 bundling of Middle Triassic Latemar cycles. The plot shows changing accommodation by graphing cumulative deviation from mean cycle thickness against cycle number (base on left) to top (on right). Insets are autocorrelation plots showing 5 to 1 bundles. From Goldhammer et al. (1990). Lower left: Graphical plots showing how carbonate cycles may be generated by fluctuating sea level (wavy line); subsidence shown by short lines sloping down to right, deposition shown by short lines sloping up to right (white areas); non-depositional intervals shown in black. Resulting columns are shown on right, with the top one being typical of the tepee facies with numerous missing beats, the middle one typical of the cyclic facies, and the bottom one typical of dominantly subtidal facies (lower platform). Lower right: Stratigraphic column and corresponding inferred 3rd order sea level curve with superimposed higher frequency fluctuations. Note how the characteristic cyclic packages relate to the 3rd order curve. From Goldhammer et al., 1990).

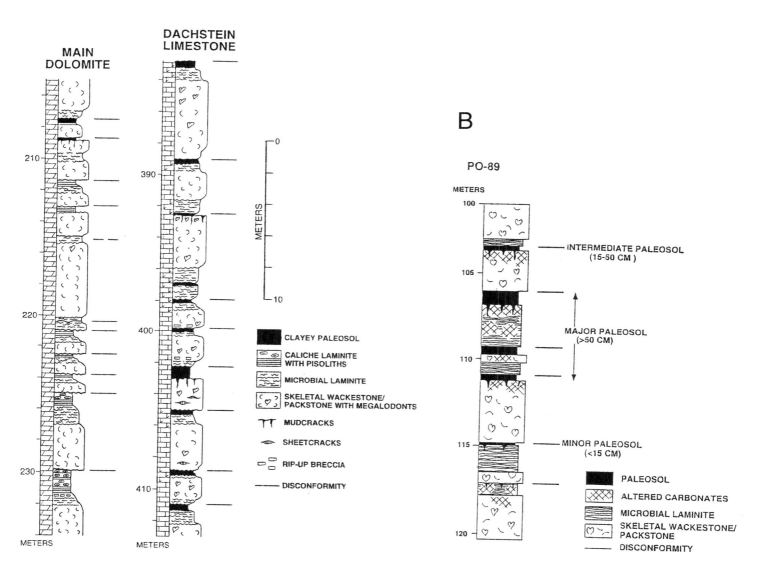

Fig. 4-15. A. Representative portion of Late Triassic stratigraphic column from Main Dolomite unit of the large Hungarian platform. Cycles have well developed laminite facies. Compare this with B. representative stratigraphic column of the overlying Dachstein Limestone unit of the platform, which shows numerous paleosols, reflecting more humid climate. The Hungarian platform shows transgressive, regressive and symmetrical cycles (Balog et al., in press).

CHAPTER 5

CARBONATE RESERVOIRS FORMED UNDER MODERATE AMPLITUDE, 4TH /5th ORDER SEA LEVEL FLUCTUATIONS

Moderate fluctuations (perhaps 20 to 50 m range) in high frequency sea level should typify times when the globe is transitional between times of major continental glaciation (commonly but not necessarily associated with icehouse conditions) and greenhouse conditions. Such times of transition might include the Late Proterozoic to Middle Cambrian, Early to Late Ordovician, the Middle to Late Devonian, the Early to Late Carboniferous, Early to Late Permian and the Tertiary to Quaternary (Fig. 1-3).

Key Features

1. Parasequences or cycles tend to be less layer cake and more shingled than low amplitude cycles.

2. Tops of platforms and inner parts of ramps tend to have more slope than greenhouse platforms

3. Deeper water tongues extend into bases of cycles on seaward parts of platforms; causes compartmentalization of shallow water bank/shoal grainstones into numerous thin units

4. Shallow water grainstones show considerable lateral migration from cycle to cycle, particularly on ramps

5. Tidal flat facies rare

6. Widespread, regionally mappable disconformities at cycle tops

7. Siliciclastic influx may be increasingly important in bases and/or tops of land-attached platform cycles

8. Moderate sea level changes may promote relatively thick (few tens of meters) meteoric systems and associated cementation that extend down through a few cycles in humid climates. In arid climates, diagenesis will be limited to calichification and possibly dolomitization of subjacent facies, although pervasive regional dolomitization may be more likely associated with progradation of late HST evaporitic tidal flats.

Sea-level Changes during Transitional Times: The stratigraphic record suggests that sea-level changes at these times show little evidence of precessional forcing. Stratigraphies suggest that sea level fluctuations are due mainly to 400 and 100 k.y. eccentricity cycles and perhaps 40 k.y. obliquity cycles (Fig. 1-5), although throughout much of the stratigraphic record it is not possible to accurately determine real periodicities because of the poor time constraints. Magnitudes of sea level fluctuations perhaps are between 20 and 50 m or so.

Platform Morphology : Tops of rimmed shelves and inner parts of ramps tend to have more slope than greenhouse platforms, because 20 to 50 m sea level changes prevent the platforms filling all the accommodation space to the 4th order highstands. The locus of maximum sedimentation rate moves rapidly landward and then seaward across the platform with sea level rise then fall. At the peak of a 4th order transgression, the outer part of the platform may be deeply submerged and sediment starved, and only undergoes rapid sedimentation as sea level falls. The shoreline rapidly regresses, successively moving the belt of maximum deposition further seaward and to lower elevations. Consequently these shelves never really reach equilibrium with respect to sea level.

Because there is unfilled accommodation following flooding of these platforms, some topography may be generated on grainstone shoals and skeletal buildups. Carbonate sand bodies may show multilateral tidal sand geometries, and multilateral stacking.

Cycle Stacking: Modelling suggests that twenty to 50 m, 100 to 400 k.y. sea level fluctuations prevents layer-cake formation of

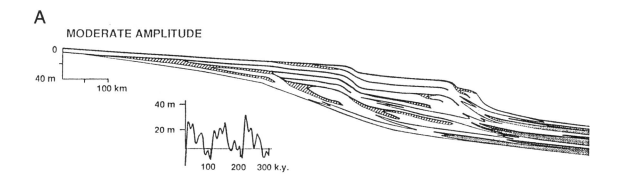
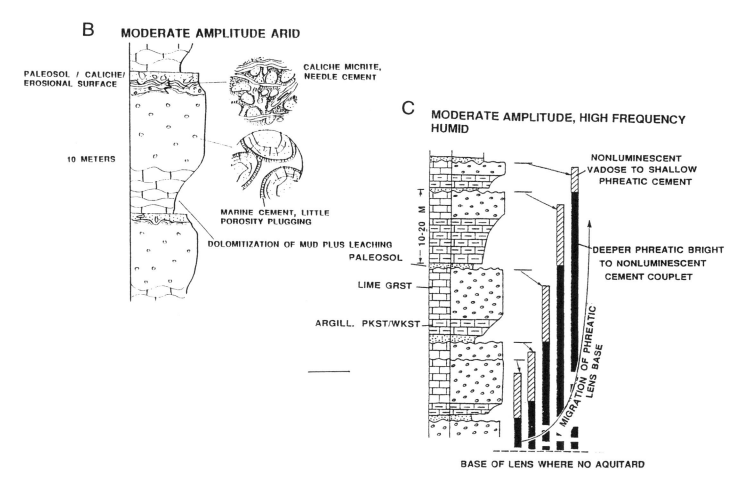

Fig. 5-1 A. Synthetic stratigraphic cross section generated under moderate amplitude Milankovitch sea level fluctuations. Under these conditions the tidal flats are not regional but are restricted to still stands of sea level, thus they commonly occur in downdip positions and pass updip into disconformities (Read et al., 1991). The small carbonate cycles are much less regional when compared to those formed under lower amplitude fluctuations of sea level.
B. Oolitic carbonate cycles formed under moderate amplitude Milankovitch sea level fluctuations and arid climate.
C. Typical skeletal carbonate cycles formed under moderate amplitude Milankovitch sea level fluctuations and humid climate (modified from Horbury and Adams, 1989). Repeated sea level fluctuations causes establishment of successive, thick meteoric lenses and their distinctive cement zones that are superimposed on earlier cements, especially in lower parts of lenses.

cycles, except on very flat topped isolated platforms. On land attached rimmed shelves and ramps, the carbonate cycles or parasequences will have a slightly shingled arrangement (Read, Osleger and Elrick, 1991)(Fig. 5-1A). Any individual precessional/obliquity cycles will be confined to narrow shore-parallel zones, whose regional position on the platform will tend to be governed by tectonic/antecedent topography, and by 100 and 400 k.y. sea level changes. That is, the 5th order cycles will tend to backstep then prograde with 100 and 400 k.y. changes in accommodation. Individual shingled 5th order cycles will not be regionally traceable downdip. The disconformity capping each shingled cycle will pass updip into a 4th order disconformity, and downdip into a conformity, and the cycle will lose its identity, merging with a 4th order cycle. In general, there are too few cycles developed on these platforms to generate reliable Fischer plots.

Facies: Outer ramp cycles show well developed moderately deep subtidal argillaceous limestone in lower parts, beneath grain-rich shallow water caps, due to rapid deepening during high frequency transgressions. These deeper water tongues die out into foreshoal facies beneath cycles onto the platform, where cycles commonly are grainstone-dominated. Significant eolian deposition may occur on arid platforms, especially updip, and at bases and/tops of cycles. On humid platforms, siliciclastics may be transported across the platform during late highstand and can accumulate along the platform margin during lowstand. Some incision of the platform may be observed. During subsequent transgression, there may be widespread deposition of marine and tidal reworked siliciclastics during backfilling of platform topography, as well as widespread eolian deposition.

Compartmentalization: Coarse grained facies of carbonate buildups and ooid grainstone shoals are highly compartmentalized by interlayered lagoonal muds updip, and by interbedded deeper water muddy carbonates downdip. This interlayering is a consequence of substantial 4th order eustatic fluctuations preventing any long term maintenance of shallow water, high energy settings over banks and shoals.

Scarcity of Tidal Flat Facies and Abundance of Disconformities Capping Cycles: Tidal flats cap only a few cycles, which instead have karstic erosional tops developed on shallow subtidal facies (Figs. 5-1B,C) (Horbury, 1987; Horbury and Adams, 1989; Wright, 1992). The scarcity of tidal flat facies on these platforms may reflect the relatively rapid fall rates during Milankovitch sea level drops. This causes the shoreline to migrate faster than tidal flats can accumulate sediment and prograde. Consequently, these platform cycles become capped by widely traceable disconformities that overlie lagoonal or grainstone shoal facies.

Stacked, offlapping tidal flat cycles can form regional units within upper parts of 3rd order HST's (Elrick and Read, 1991), perhaps reflecting limited accommodation and relatively low amplitude sea level fluctuations during 3rd order highstand, compared to the moderate amplitude eustasy of the TST's. This situation is most likely to occur where high frequency amplitudes are still relatively subdued. It does not appear to occur where the sea level fluctuations are larger.

Linear belts of tidal flat facies up to a few meters thick may occur in both updip and downdip parts of some platforms during 4th and 5th order highstands and lowstands of sea level (Fig. 5-1A). At these times the near-zero, 4th and 5th order eustatic sealevel change probably favors development of tidal flats because the shoreline is relatively static.

The better developed disconformities that commonly cap cycles on these platforms, compared to greenhouse platforms, reflect the longer duration of exposure of each cycle. If most of these platforms have cycles dominated by 40, 100 and 400 k.y., then exposure times may be tens to a few hundred thousands of years. In addition, the larger sea level changes may cause considerable lowering of groundwater tables during lowered sea levels, favoring intense vadose diagenesis beneath vegetated cycle tops. Disconformities die out into deeper ramp/foreslope settings, which are below the position of 4th order low stands.

Diagenesis and Reservoirs: Depositional porosities are highest in grainy upper parts of cycles, and porosity zones tend to be thicker (up to 10 m or more) than greenhouse cycles. Also, argillaceous, muddy lower parts of cycles tend to act as partial internal seals resulting in subseismic stratified reservoirs.

Arid zone carbonate cycles (Fig. 5-1B) can have porosity in tops of cycles plugged by caliche formation and vadose fibrous cements. Subtidal grainy parts of cycles lack early sparry cements and retain much primary porosity especially where they have marine fringe cements. Regionally prograding peritidal dolomites and evaporites can form seals to 3rd order sequences.

Humid zone cycles (Fig. 5-1C) may show some karsting and soil formation at cycle tops, while upper parts of cycles undergo some moldic and vuggy dissolution, along with plugging of primary and secondary porosity by vadose and upper phreatic sparry calcite cements. Middle or lower parts of cycles, if grainy, may be less well cemented and retain more primary porosity. Muddy cycle bases undergo stabilization of carbonate muds in meteoric waters and some moldic porosity formation. Because of development of thick ground water zones, meteoric diagenesis may extend down into underlying older cycles, if seals are poorly developed. Sub-seismic pre-burial reservoirs will be stratified, with relatively homogeneous porosity and some permeability baffles, and should be identifiable from cores and electric logs.

Example 1: Early Mississippian Reservoirs of Wyoming-Montana:

The Early Mississippian Madison Group in the U.S. Rocky Mountains predominantly formed under low to moderate amplitude sea level fluctuations (Elrick and Read, 1991), perhaps interspersed with times of lower amplitude fluctuations during 3rd order highstands. Note that the magnitude of these sea level fluctuations likely were lower than those of the later Missisippian in the Appalachians. The Madison Group consists of a major 2nd order sequence, composed of basin/deep ramp to shallow ramp to tidal flat/sabkha facies and solution collapse breccia (Lodgepole-Mission Canyon-Charles Evaporites). The major sequence has several 3rd order sequences with well developed 1 to 10 m thick parasequences or cycles (Dorobek, pers. comm., 1991); Elrick and Read, 1991) (Fig. 5-2).

Transgressive systems tract cycles in the Lodgepole Formation consist of commonly dolomitized skeletal wackestone/mudstone passing up into ooid grainstone with little evidence of disconformity development. Downslope, the cycles are deeper water, laminated or burrowed dark muds that grade up into storm-deposited skeletal/oolitic packstone-grainstone caps. Tidal flat facies are generally absent. In contrast, high stand systems tract cycles are dominated by dolomitized thin peritidal cycles of subtidal oolitic/pelletal carbonates overlain by tidal flat laminated caps that show brecciated tops near sequence boundaries updip.

Modelling of the cycles suggested that the parasequences were formed by low-moderate (less than 40 m) fluctuations in sea level, in order to generate transgressive systems tract, oolitic cycles free from tidal flat facies on the shallow ramp while at the same time deeper water mud up into storm-deposited grainstone cycles were forming on the deep ramp/slope. However, the high frequency sea level fluctuations needed to be reduced in magnitude in order to generate the well developed tidal flat capped cycles of the high stand systems tract. This raises the likelihood of the high frequency sea level fluctuations being highest in the transgressive systems tracts, decreasing into the high stand systems tracts.

The Whitney Canyon-Carter Field, Wyoming, is in the Madison Group and is volumetrically the largest gas producer in the U.S. Rocky Mountains (Harris, Flynn and Sieverding, 1988). Most production is from over 100 m of porous fine dolomite interbedded with fine calcitic dolomite and tight lime grainstone (Fig. 5-3). This porosity zone is underlain by fine grained, deeper water basinal limestone and overlain by a fractured and brecciated section of sabkha limestone and dolomite. In the main

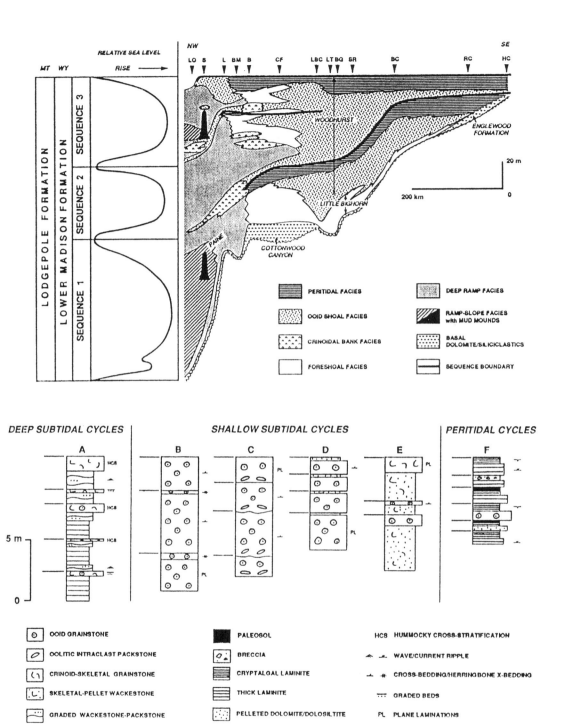

Fig. 5-2. Top: regional cross section approx. 700 km long, showing depositional sequences in the Mississippian Lodgepole Formation, believed to have developed under high frequency, moderate amplitude eustasy (Elrick and Read, 1991). Bottom: examples of cyclic carbonates across the ramp. Note the relative scarcity of tidal flat facies except within the HST, the stacked shoals on the mid-ramp, and the coarsening upwards units further downslope which resemble some of the reservoirs in the subsurface (Elrick and Read, 1991).

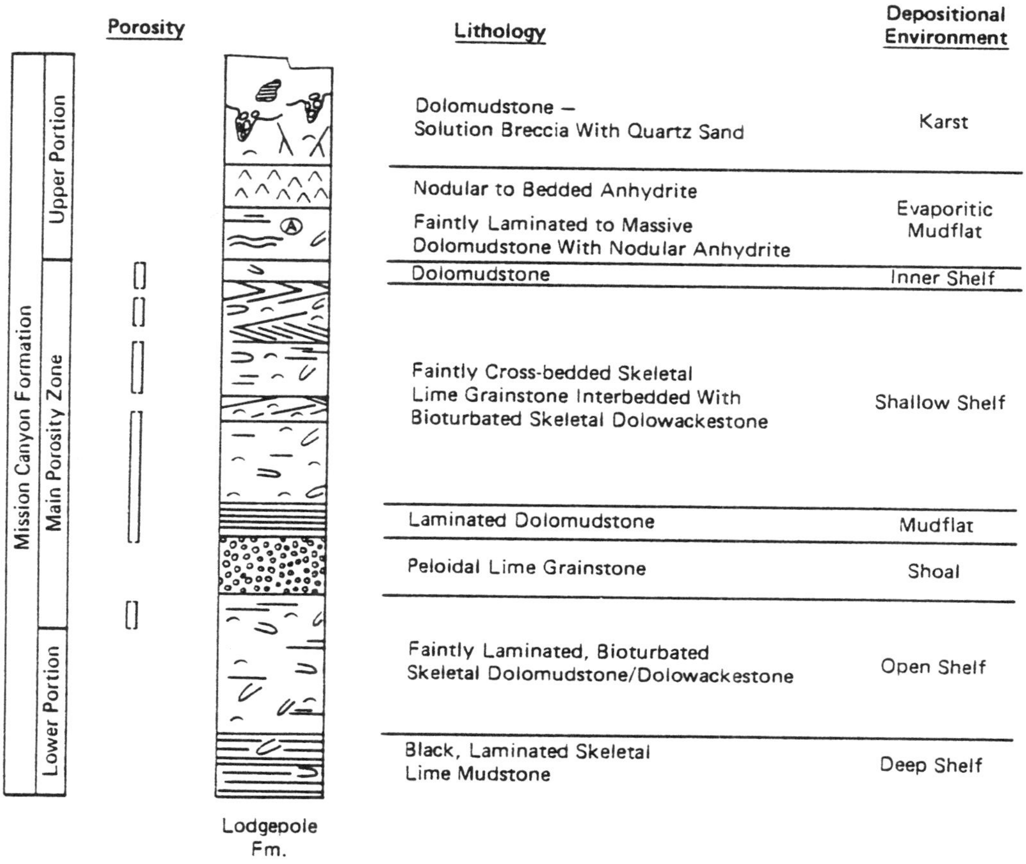

Fig. 5-3. Idealized composite section of Mississippian Mission Canyon in Whitney Canyon-Carter field, Wyoming (Harris, Flynn and Sieverding, 1988).

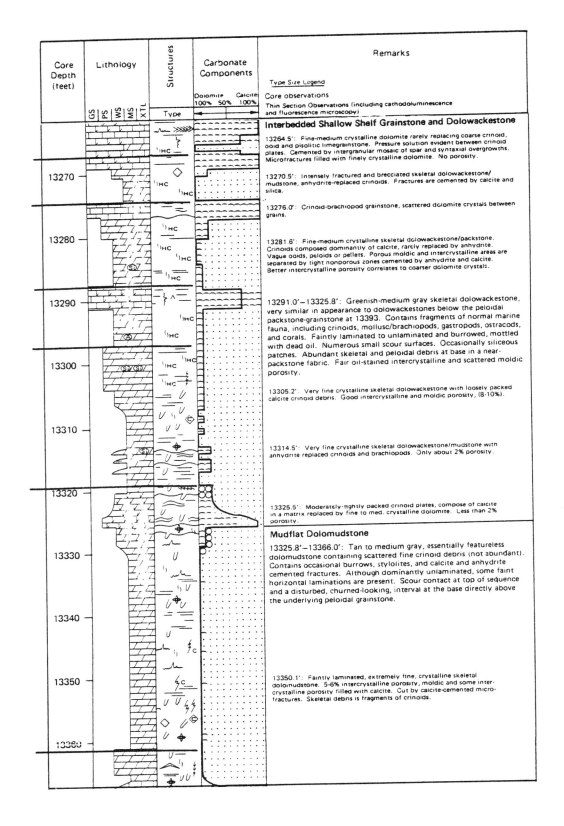

Fig. 5-4. Example of core description of Mississippian rocks from Whitney Canyon-Carter field, Wyoming, showing inferred shallowing- and coarsening-upward cycles in the dolostones (from Harris, Flynn and Sieverding, 1988).

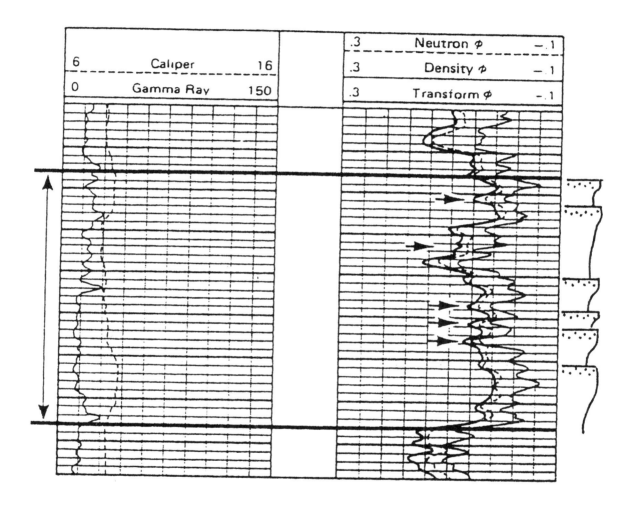

Fig. 5-5. Well logs for the producing interval, Mississippian Whitney Canyon-Carter field, Wyoming. Arrows indicate porous subtidal muddy carbonates, whereas tight intervals are grainstone rich, illustrated by schematic log on right. Modified from Harris et al., 1988.

porosity zone, the best reservoirs occur in cycles of porous fine grained, dolomitized subtidal wackestone/mudstone overlain by tight crinoidal grainstone (Figs. 5-4 and 5-5). The fine to very fine grained dolomite forms the reservoir facies and has intercrystal porosity along with skel-moldic porosity. The dolomites likely have complex origins and may have formed from hypersaline waters during intermittent shallowing, from mixed waters, or from brines from the overlying, evaporitic sabkha facies in the upper Madison Group. These brines refluxed downsection to dolomitize permeable subtidal wackestone/mudstone but not the cemented (?) grainstone units which now form tight limestone units. The Mission Canyon dolomites subsequently were modified in meteoric or mixed water environments recharged from the overlying unconformity and possibly reset during burial (Smith and Dorobek, 1993; Dorobek et al., 1993). The tight lime grainstones may have escaped dolomitization because they were already cemented (Harris, Flynn and Sieverding, 1988), although Dorobek et al. (1993) considered the much of the Mission Canyon calcite cements post-date the early dolomite.

Example 2: Arid Zone Ramp Reservoirs, Late Mississippian, Eastern and Central U.S.A.
(A. Al-Tawil and L.B. Smith):

The Late Mississippian carbonates and clastics in the eastern U.S. appear to have formed during transition into the Pennsylvanian glaciation, and amplitudes appear to have been higher than in the earlier Mississippian. These dominantly fourth order cycles are probably 100 to 400 k.y. duration, which formed on a broad ramp (Figs. 5-6 to 5-8). They include oolitic and skeletal grainstone cycles up to 10 m thick, capped by microkarsted surfaces, caliches and detrital chert, and rare sandy/silty dolomites (Niemann and Read, 1988; A. Al-Tawil and L.B. Smith, unpublished data). The disconformities capping cycles are marked by paleosols of coated 'dirty' grainstone/packstone (incipient soil), caliche stringers in slightly brecciated host (incipient caliche) or heavily calichified host, with locally intense brecciation and tepee development and paleosinkholes forming an undulating contact (mature exposure surfaces) (Ettensohn et al., 1988). Many of these paleosols are regionally traceable over much of the platform (Figs. 5-6 and 5-7). Caliches and karsting are best developed at the 3rd order sequence boundaries. Updip facies include redbeds and quartz-carbonate eolianites (Hunter, 1993). Lagoonal muds and rare tidal flat laminites occur either below or above ooid or skeletal facies in cycles. Oolitic bodies are both parallel and normal to paleostrike, and appear to form discrete bar form carbonate sand bodies interpreted as tidal bars (Kelleher and Smosna, 1993). Lagoonal and rare tidal flat facies are relatively rare capping cycles, but may form thick units during 4th order late highstand and lowstands on the ramp. Tectonics interacting with eustasy has profoundly controlled the regional distribution of grainy facies and disconformities.

Facies do not form simple upward-shallowing cycles. Some are clearly transgressive cycles capped by a disconformity. Others are regressive cycles capped by a disconformity, and some show transgressive-regressive stacking and a capping disconformity. Siliciclastics may occur in upper parts of some cycles associated with mobile grain shoals, tidal inlets and flats (Handford, 1988) and they help divide the section into cyclic units using gamma ray logs (Fig. 5-6). However, in later Mississippian sections, where amplitudes of sea level fluctuations may have been high, clastics also are localized within transgressive parts of cycles, where they occur as marine reworked tidal-esturaine and marine sands/shales, overlain by 4th order high stand oolitic/skeletal carbonates (L. B. Smith, pers. comm., 1994).

The semi-arid climate inhibited diagenetic alteration during the moderate amplitude fluctuations because meteoric ground waters were poorly developed. Subaerial diagenesis was mainly restricted to upper parts of cycles where caliche crusts and detrital carbonate- and chert-veneered erosion surfaces formed (Niemann and Read, 1988; Harrison and Steinen, 1978). The main cements within the caliche profiles are fibrous turbid cements

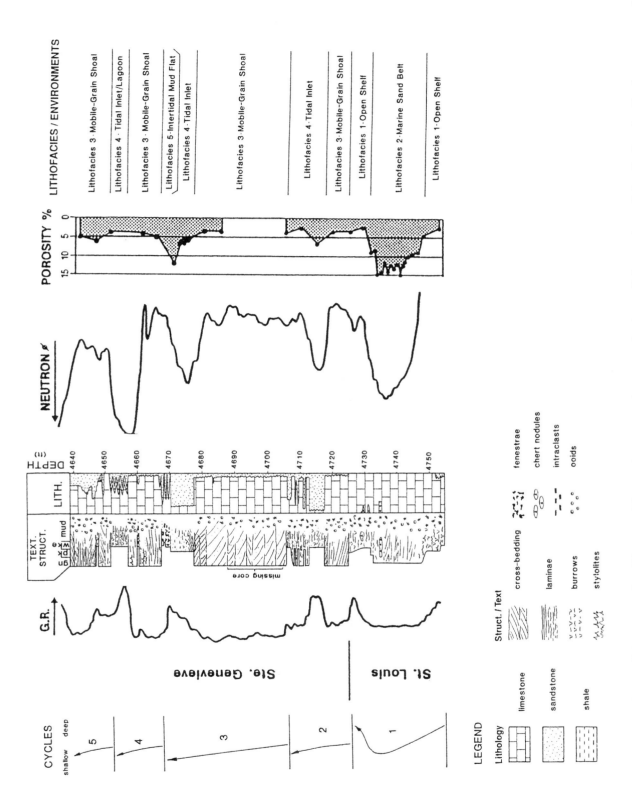

Fig. 5-6. Mississippian oolitic reservoirs, Kansas. The major reservoirs occur in the oolitic portion of the St. Louis. Cycles are grossly delineated by the gamma ray logs where they contain a clastic component. Modified from Handford, 1988.

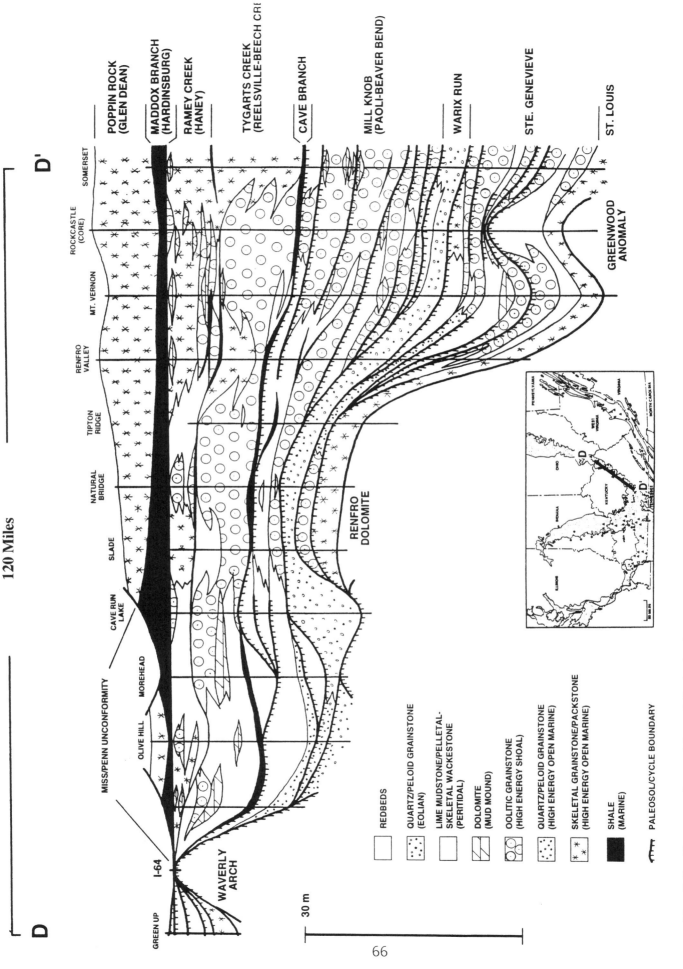

Fig. 5-7. Cross-section of Late Mississippian ramp carbonates in central Kentucky; St Louis to Glen Dean interval (Aus Al-Tawil, unpublished data). The rapid thickness variations are due to tectonics, which interacted with 4th order eustasy to generate the disconformity bounded cycles. Note the downdip increas in number of cycles, and stacking of disonformities over highs.

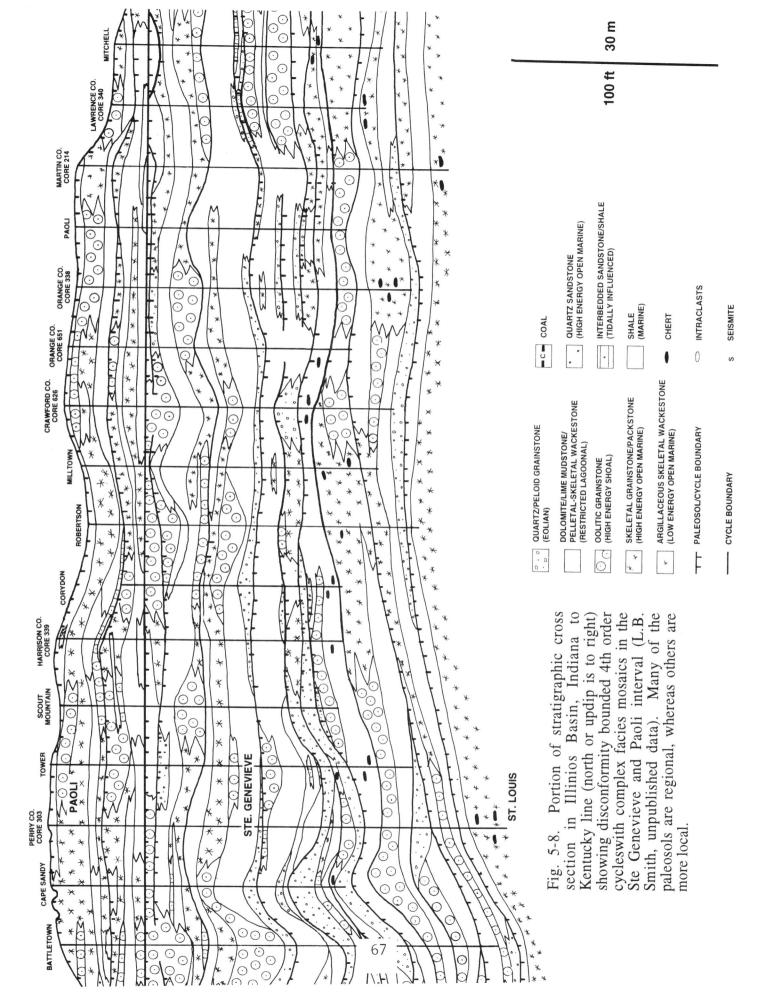

Fig. 5-8. Portion of stratigraphic cross section in Illinios Basin, Indiana to Kentucky line (north or updip is to right) showing disconformity bounded 4th order cycleswith complex facies mosaics in the Ste Genevieve and Paoli interval (L.B. Smith, unpublished data). Many of the paleosols are regional, whereas others are more local.

(needle fiber calcites), micrite coatings on grains and rhizocretions formed by carbonate precipitation on roots and rootlets (Fig. 5-1B); sparry calcite cementation and leaching was relatively minor until the Late Mississippian-Early Pennsylvanian when the climate became humid. Because there was little early sparry calcite cementation of oolitic grainstone facies during 4th/5th order sea level falls, the distribution of the hydrocarbon reservoirs is tied to the lateral and vertical distribution of grainstone bodies (Handford, 1988; Kelleher and Smosna, 1993)(Fig. 5-6). Intergranular porosity was decreased in more deeply buried downdip sections by over-compaction during burial. Chalky microporosity is important in some of the gas reservoirs in the oolite units, although it is not clear whether this relates simply to later emergence and sequence-bounding unconformity development or is due to burial fluids. Dolomitization, either by hypersaline brines or mixed waters, and in some cases by burial fluids, has modified muddy carbonates in the cycles, locally forming microporous reservoir zones (Choquette and Steinen, 1980).

Example 3: Humid Zone Platforms British Dinantian (Mississippian) (modified from A. Horbury):

Northern British Mississippian cycles were formed on isolated- and land-attached rimmed shelves (Somerville, 1979a,b; Walkden, 1987; Horbury, 1989). Cycles are capped by 30 or more well developed paleokarstic surfaces (Fig. 5-1C) and have periodicities of 100 to 200 k.y. and are related to onset of Permo-Carboniferous Gondwana glaciation. Cycles are made up of skeletal and peloidal packstone and grainstone (and rare oolite and fenestral mudstones) in upwards shoaling units. Regression was rapid as indicated by subaerial exposure surfaces and paleosols directly on subtidal facies (Horbury, 1989; Walkden and Walkden, 1990). Paleosol fabrics are complex (Davies, 1991), and reflect a dry-wet-dry cycle (S. Vanstone, pers. comm. 1992), which is probably related to 5th order climatic cyclicity superimposed on 4th order climate cycles (Gray, 1981, quoted by Tucker, 1985).

They comprise a thin, rooted rhizoconcretionary interval up to 1 m thick (semi-arid phase), cut by a clay-covered, mammilated paleokarst with up to 2 m relief (wet phase) (Walkden, 1987; Horbury, 1989). Superimposed later fabrics such as laminated caliche and breccia (semi-arid phase) are rare.

In cycles on the isolated rimmed platforms, the shallowest subsurface meteoric cements are rare pendant and ponded vadose brown fibrous cements, and brown equant cements which are very dull luminescent and related to calcrete mottle formation in soil zones. These are the dry climatic phase and were short-lived rapid precipitation events in localized diagenetic environments (Horbury, 1987; Horbury and Adams, 1989). There also are shallow (5 to 15m depth) meteoric cements, which formed by reprecipitation of the bulk of the calcite dissolved beneath the paleokarst, probably during the wet climatic phase. These cements are volumetrically abundant and are clear nonluminescent calcites with very fine bright subzones which cannot be correlated for more than a few millimeters. They develop on all substrates but are thicker as syntaxial cements and are best developed close (5 to 15m) beneath the paleokarst from which they are sourced (Horbury and Adams, 1989). In contrast, deeper in the phreatic lenses, well developed correlatable zoned clear calcite cements occur (Fig. 5-1C). These are a volumetrically insignificant but important cement type that formed by slow precipitation of calcite in deeper (15 to 120 m subpaleokarst) parts of meteoric lenses of successively younger paleoaquifers. These cements only occur where rimmed shelves are relatively isolated from probable laterally-sourced meteoric water, e.g. areas of permanent emergence such as onlap margins onto basement. Each emergent event was recorded by growth of a couplet of bright to non-luminescent cement. Hence there are roughly the same number of cement couplets as there are eustatic cycles containing these cements (Fig. 5-1C). Where meteoric lenses rested on a shale aquitard, the cement couplets are most abundant low in the section, and decrease upward into younger cycles. They extend over a limited vertical range beneath the parent paleokarst surface, with younger couplets progressively appearing upsection (Horbury and Adams, 1989).

Example 4: Later Ordovician, Kentucky, USA (Mike Pope):

The late Middle Ordovician Lexington Limestone of Kentucky, U.S.A. (Fig. 5-9) has features suggestive of moderate amplitude eustasy on a cool water ramp. The Lexington Limestone is the basal 3rd order sequence of the thick Lexington to Cincinnatian 2nd order depositional sequence (lower Trenton Group). The ramp sloped gently north and east from a forebulge (Cincinnati Arch) into the Appalachian foreland basin and passed abruptly westward into the narrow Sebree Trough.

The Lexington to Cincinnatian units are dominated by subtidal facies, with rare tidal flat- and skeletal shoal-facies developed over the arch during shoaling. Facies (Fig. 5-10) are: 1) inner ramp tidal flat laminated/fenestral lime mudstone/restricted skeletal wackestone, with karstic features and early meteoric diagenetic features near the sequence boundary; 2) lagoonal nodular-irregularly bedded whole-skeletal wackestone/packstone with stromatoporoids, corals, pelecypods, gastropods, and red algae ; 3) mid-ramp tidally influenced skeletal grainstone/packstone sheets or banks; and 4) deeper ramp nodular-irregularly bedded skeletal wackestone/packstone (bryzoan-echinoderm-brachiopod) and shale grading seaward into regular bedded shale and limestone storm-beds; and in sheltered low energy environments behind and rarely in front of the banks, "basinal" evenly thin-bedded calcisiltite and shale.

Facies commonly are arranged in shallowing-upward, dominantly subtidal cycles (Fig. 5-10), which are asymmetric to symmetric, 1-7 m thick. Most are of thin-bedded/nodular shale and skeletal limestone, capped by tidal/storm-deposited skeletal grainstone topped by pyritized hardground(s), and some contain a basal unit of dark, thin bedded calcistiltite and shale. Based on available chronostratigraphy, cycles are 40 to 130 k.y. duration. These cycles are arranged into asymmetric cycle-sets or bundles (possibly around 1 m.y. duration) marked by basal regional flooding units. Cycles with "basinal' thin bedded shaly limestones grading up into storm/tidal grainstones appear to have formed during moderate sea level changes (few tens of meters) whereas the bulk of cycles lacking juxtaposed deep- and shallow water facies probably formed under smaller sea level changes. The change from shale-rich up into carbonate-rich units in cycles, could also be reflecting change from wet to drier climate during cycle deposition.

Tidal flat cycles occur in 3rd order late highstands/lowstands and are suggestive of smaller sea level fluctuations than TST's, because few subaerial emergence features are evident in the peritidal beds, except along sequence boundaries.

Grainstone/packstone units of the Tanglewood bank facies are compartmentalized by numerous dark, deeper water, muddy beds which partition the bank (Fig. 5-9). This partitioning of the bank facies likely is a signature of moderate (and high) amplitude eustasy.

The abundant brachiopods and bryozoans, presence of low-Mg calcite marine cements, iron and phosphatic stained hardgrounds, and the low accumulation rates (1.4 to 2.5 cm/k.y.) and lack of any warm-water chlorozoan fauna, ooids, or evaporites, indicate dominantly cool water deposition. Stromatoporoid and coral horizons in the late highstand may indicate warmer conditions. The moderate amplitude sea level fluctuations indicated in the subtidal cycles may be due to glacio-eustatic sea level changes associated with waxing and waning of incipient Gondwana ice sheets during a time of global greenhouse climate. Furthermore, it appears that cool marine bottom waters in the highly stratified epeiric sea were present to paleolatitudes of 10-30° S during the late Middle Ordovician

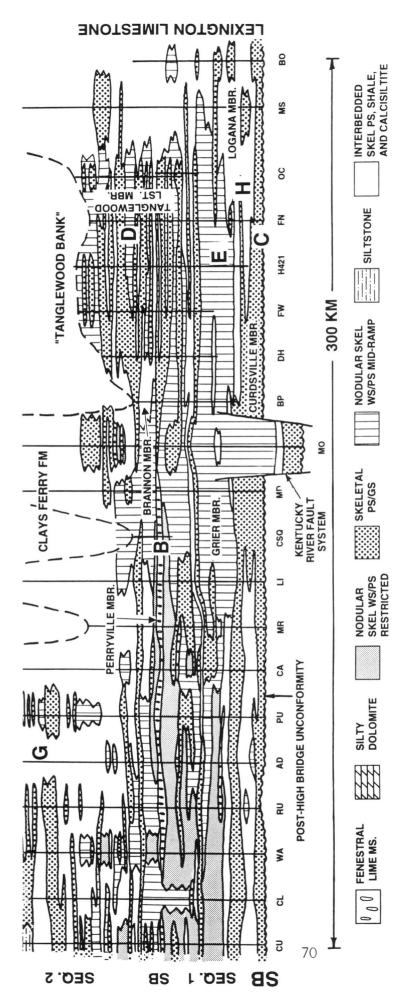

Fig. 5-9. Regional S-N cross section of Late Middle Ordovician Lexington Limestone (Sequence 1 in figure) along Cincinnati Arch, Kentucky (M. Pope, in press). Note the compartmentalization in the Tanglewood bank.

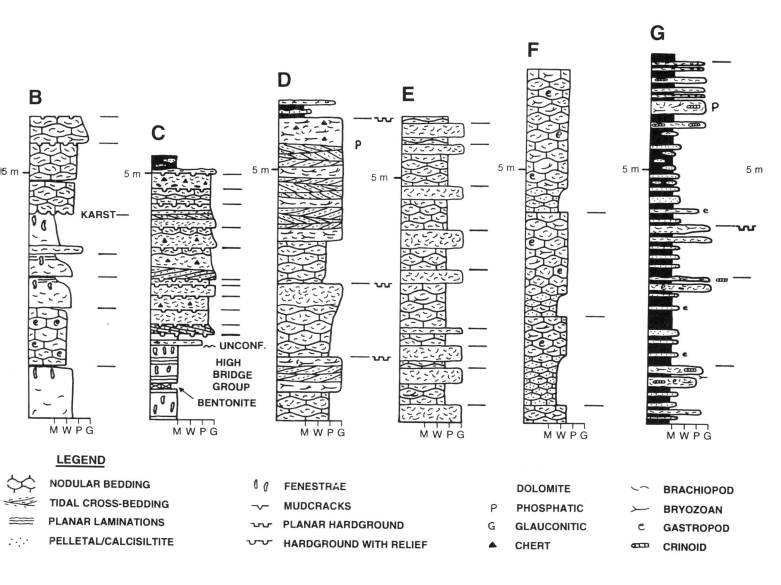

Fig. 5-10. Portions of stratigraphic columns showing typical cycle types arranged from shallow to deep ramp, Lexington Limestone and overlying units, Kentucky (M.Pope, in press). Cycles appear to reflect moderate amplitude eustasy on a cool water ramp over a tectonically active arch.

CHAPTER 6

CARBONATE RESERVOIRS FORMED BY HIGH AMPLITUDE 4TH/5TH ORDER GLACIO-EUSTASY

Large, high frequency fluctuations in sea level (60 to over 100 m oscillations) appear to have characterized global icehouse times of the Pleistocene, the Pennsylvanian-Early Permian and the Late Precambrian (Fischer, 1982)(Fig. 1-3).

1. Parasequences or cycles dominated by eccentricity (100 to 400 k.y.) rhythms or obliquity rhythms

2. Platform tops sloping, reaching a meter/km on ramps

3. Flat-topped platforms show layer cake, 1 to 10 m thick, dominantly subtidal disconformity-bounded cycles . Along margin, reef crest facies show tens of meters of vertical shift during 100 to 400 k.y. sea level changes.

4. On ramps, cycles are erosionally bounded, and have highly shingled stacking pattern,

5. High relief, pinnacle reefs and banks are common on tropical platforms

6. Cycles lack tidal flat facies, except adjacent to shoreline. Most cycles have regional disconformities

7. Except on the shallowest parts of platforms, cycles will show juxtaposition of deep water facies, shallow water facies and subaerial emergence features, indicating large sea level changes

8. Large sea level fluctuations cause very large vertical and lateral migration of diagenetic zones and groundwater tables. Karstic sinkholes and cave systems may extend down through several cycles. Still-stands may localize paleowater tables repeatedly at various levels in the stratigraphy, causing intense leaching/cementation and perhaps dolomitization.

Sea-Levels During Times of Major Continental Glaciation:

During times of major continental glaciations, sea level changes are likely to be large (60 to 100 m or more), with periodicities dominated by 100 and 400 k.y. as in the last 700 k.y., or by 40 k.y. obliquity cycle which typifies the earlier Pleistocene record (Fig. 1-5) (Ruddiman and Wright, 1987). Sea level rise is rapid (at least several meters/k.y. and possibly even faster), and regressions probably are slower. Precessional cycles are overshadowed by these longer term cycles. However, higher frequency glaciation and deglaciation a few k.y. to 13 k.y. duration and below the Milankovitch band also may be important.

Platform Morphologies:

Because sea level changes are large, platforms rarely attain equilibrium with sea-level. Consequently the tops of ice-house platforms tend to be much steeper than other times. Ramps will tend to slope from the post-glacial highstand elevation down to 60 to 100 m or more below this, unless modified by underlying topography/structure. Interiors of rimmed margins can have slopes of a few cms/km (Bahamas), with steep marginal slopes. Rapid sea level rise favors formation of elevated reefal rims, pinnacle and platform reefs, and these may be backed or surrounded by deep lagoons. The buildups are due to high sedimentation rates on reefs and banks which tend to track the rapid sea level rise, while the rest of the shelf tends to drown. Margin facies show successive, large upward and downward shifts of prograding reef crest facies, tracing the large scale (tens of meters) vertical shifts in sea level.

Cycle Stacking:

Cycles on rimmed platforms are disconformity-bounded, layer-cake, 1 to 10 m units, but these become highly shingled along the margin (Fig. 6-1B). On ramps, cycles are highly imbricated with any small,

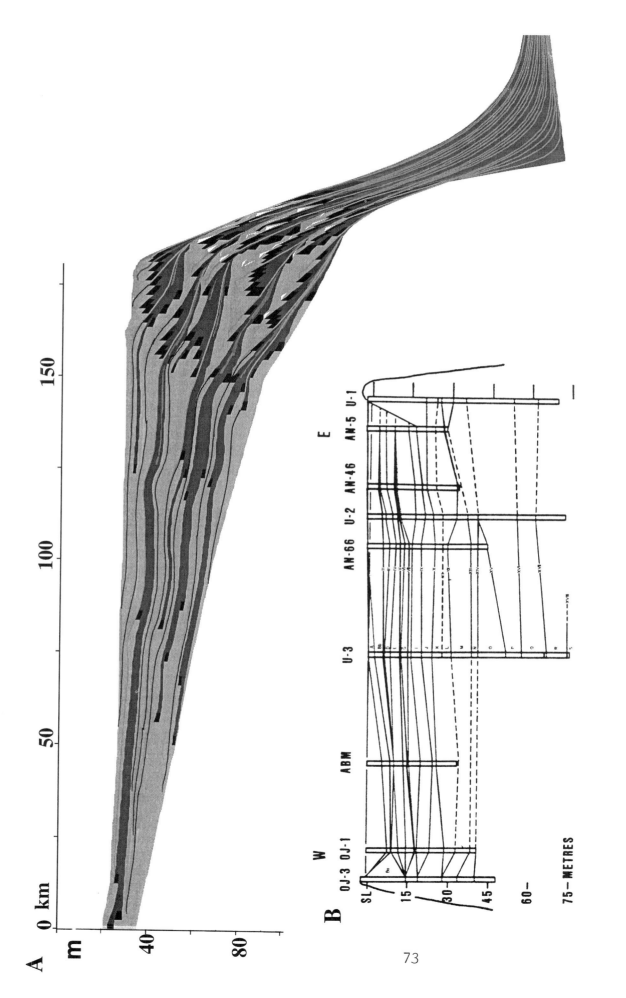

Fig. 6-1 A. Synthetic stratigraphy made using Scott Bowman's PHIL on a platform that developed under high amplitude Milankovitch sea level fluctuations. Cycles are shingled, and show well developed low stand wedges, and deeper water facies extending onto platform.
B. Correlation of subaerial discontinuity zones across northeastern Great. Bahama Bank (Beach 1982).

geographically restricted 5th order cycles backstepping then offlapping, within 100 to 400 k.y. depositional sequences/parasequences (Read, Osleger and Elrick, 1991). At any individual section, the major disconformity bounded cycles will be dominantly 4th order, with any higher frequency, small scale cycles shingled laterally across the platform rather than being vertically stacked.

Individual cycles may show considerable thickness variation, reflecting the large amount of unfilled accommodation over much of the platform, with local shoaling associated with high sediment production/accumulation on reefs, banks and ooid shoals. Such contour parallel "thicks" are most likely to form during high-frequency still stands during transgression, but they could also form during still-stands during 4th order progradational phases. During stillstands, sediment is eroded from updip areas (wave planated terrace) and deposited immediately downdip, to form regional terraces on the shelf.

Cycle Facies:
On parts of the platform with sufficient accommodation, individual cycles show juxtaposition of very deep water facies (spiculite, pelagic carbonates, phosphatic black shale), shallow subtidal facies, and subaerial exposure features (karst, paleosols). The 1 to 10 m cycle thicknesses are far too small to account for this juxtaposition by upward shallowing by sedimentation, which requires many tens of meters of sea level change. Cycles may be symmetrical transgressive-regressive over much of the outer platform/ramp, but on the inner ramp they commonly are asymmetrical, upward shallowing. Lowstands on ramps free of siliciclastic influx may be marked by burrow-homogenized deep water pelagics and low-stand shallow water sediment such as ooids (Logan et al., 1969).

Widespread marine to near-shore clastics may be deposited on the shelf during late highstand, followed by incision, and lowstand shelf edge deltaic deposition along the margin. During transgression, sediments are reworked into transgressive lithoclastic or quartzose veneers.

Tidal flat facies are rare on ice-house platforms because sea level falls off the platform far faster than tidal flats can prograde. Thus any high-stand tidal flat facies and associated dolomites are likely to be in narrow bands adjacent to cratonic shorelines or in the lee of cemented islands and form toward the 4th order highstand, when rate of sea level change is low.

Disconformities:
Disconformities are very well developed capping cycles, with intense weathering features, paleosol formation and calichification and/or karsting depending on climate (Perkins, 1977; Beach, 1982)(Fig. 6-1A). Sinkholes may extend down through several cycles to paleowater tables. Disconformities on 5th order cycles are local, pinching out a short distance downdip and merging updip with regional disconformities associated with 4th order sequences. These regional disconformities will extend some distance down the foreslope. On land-attached platforms, rivers may deeply incise into the platform during late highstand/lowstand. Interfluve areas form highs which may localize reefs during subsequent marine transgression, while valleys ultimately become filled with finer deeper lagoon marine sediment. Antecedent topography associated with buildups from the previous cycle will tend to localize subsequent buildup development.

Diagenesis and Reservoirs:
High amplitude sea level fluctuations are likely to leave diagenetic signatures reflecting the large scale vertical and lateral migration of diagenetic zones. These cause the rocks to be subjected repeatedly to meteoric vadose, meteoric phreatic, mixing zone and sea water diagenetic environments (Fig. 6-2). Any meteoric diagenesis during 4th or 5th order highstands would be restricted to eolian islands or newly emergent shoals where sediments can undergo cementation and leaching of aragonite associated with island water tables that may be ephemeral (Budd and Land, 1990). However the bulk of the diagenesis is likely to occur during regional platform emergence, associated with lowered sea levels during 20, 40, 100 or 400 k.y. still-stands.

Humid zone carbonate cycles can have

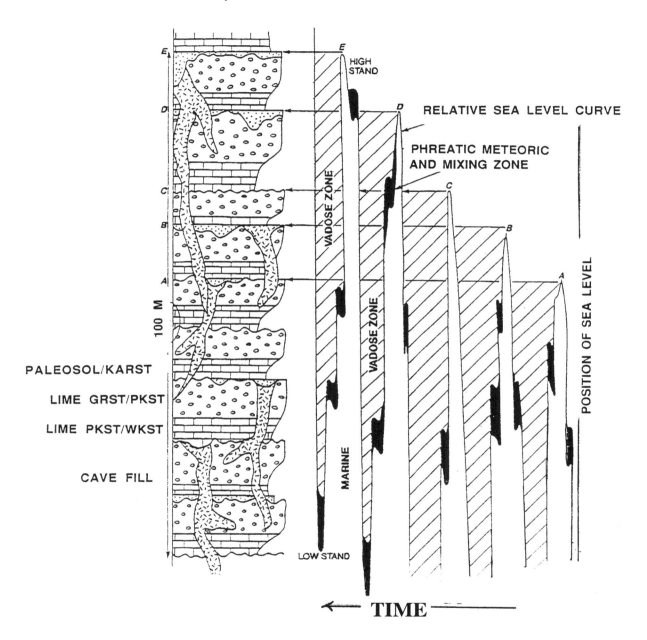

Fig. 6-2. Diagenetic model of carbonate cycles that formed under high frequency, high amplitude sea level fluctuations. Cycles are disconformity-bounded, may be karsted with cavern systems extending down through several cycles, and sediments are subjected to a complex sequence of diagenetic fluids and events, due to the large scale sea level fluctuations. Modified from Matthews and Frohlich (1987).

single to multiple caliche horizons capping cycles if there is wet-dry seasonality and numerous karstic caverns and sinkholes can extend downward through several cycles. Cycles that contain internal seals of deep water shale/carbonate can have primary porosity preserved in early cemented and leached, regressive upper parts of cycles that resist compaction (Heckel, 1983) to form highly stratified reservoirs. Coarser grained cyclic successions such as high energy reefs may lack internal seals and can form poorly stratified reservoirs. In successions of high amplitude carbonate cycles with poorly developed internal seals, uppermost cycles can have primary intergranular porosity, while lower ones will show increasing moldic, vuggy and cavernous porosity due to superimposed meteoric diagenetic events caused by large scale sea level fluctuations. It may be difficult to identify these subseismic scale cycles in core or wireline logs because of the extensive diagenetic overprint.

In arid zone cycles formed under large, high frequency sea level fluctuations, porosity may be plugged at cycle tops by caliche, but below this zone, porosity will be mainly primary intergranular in non-dolomitized buildup and shoal-grainstone units. In dolomitized cycles related to high stand sabkas on the inner platform, or late high stand to low stand evaporite basins whose refluxing brines dolomitize underlying facies, the porosity will be secondary intercrystal and remnant primary intergranular porosity in the dolomites along with primary intergranular porosity in limestone.

Example 1: High Amplitude Shelf Edge Cycles, Late Miocene Reefs, Spain:

The detailed geometries along an exposed rimmed shelf margin that formed under high amplitude sea level oscillations (Fig 6-3) have been described by Pomar (1991 and in press). The platform consists of bedded lagoonal facies, reef core, reef slope and open shelf facies. The facies together make up sigmoidal units, whose tops are erosionally bounded updip and pass downdip into conformities. These are bundled into sets of sigmoids, composed of a lower prograding unit, an aggrading unit, an upper prograding unit that downlaps onto a condensed interval, and an offlapping/downstepping unit. The set of sigmoids has an erosional top that correlates downdip with a conformable contact; perhaps each set of sigmoids represents a precessional cycle although this is not clear. These sets of sigmoids are in turn bundled into cosets, each containing a lower progradational, an aggradational, upper progradational and a downstepping/offlapping package and an upper major erosional surface; each coset is considered by Pomar (1991) to be a 100 k.y. cycle. These probable 100 k.y. cosets are in turn bundled into possible 400 k.y. packages themselves characterized by progradation to aggradation to progradational and downstepping geometries. Finally, the prograding reefs form the HST of one of the Late Miocene 3rd order sea level cycles. The reef-rimmed platform was prograding into over 100 m of water, and the 100 k.y. sea level changes likely were 60 to 70 m whereas the higher frequency ones were about 10 to 30 m.

Example 2: Pleistocene Cycles and Humid Zone Diagenesis, Bahamas:

Bahamian Pleistocene platform cycles are 1 to 10 m thick, disconformity bounded carbonates (Beach, 1982) (Fig. 6-1B). Karst features, caverns, vugs, leached zones occur at many levels (Beach, 1982; Dill, 1977; Ginsburg et al., 1990). Such karst features include large sea water-filled sinkholes over 100 m deep, extensive solution-enlarged joints (hundreds of meters to kilometers long) parallel and perpendicular to the bank margin, and vertical and horizontal passages commonly many meters wide (Dill, 1977; Smart et al., 1988). These are enlarged by meteoric dissolution at the surface, and in the shallow subsurface by waters in the mixing zone which is up to 10 m thick. They may have partial to complete fills of blocks and finer sediment of marine and terrestrial origin. These cavernous zones honeycomb the platform (Dill, 1977; Beach, 1982; Ginsburg et al., 1990).

In Bahamian platform cores (Beach,

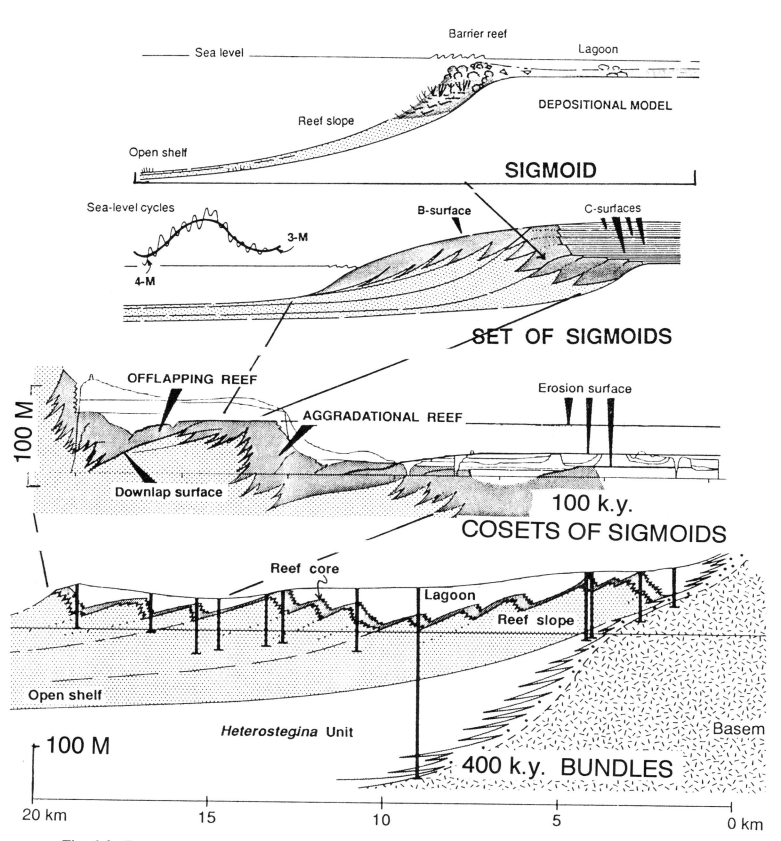

Fig. 6-3. Response of reef rimmed shelf margin to high frequency high amplitude sea level fluctuations, Miocene of Spain (from Pomar, 1991 and in press). Basic unit of the shelf margin is the sigmoid (top). These sigmoids are arranged into sets (next diagram) and then into cosets of sigmoids, considered to be 100 k.y. cycles (2nd diagram from bottom). Finally, these are bundled into larger scale packages equated with 400 k.y. cycles (bottom).

1982), the complex interlayering of diagenetic phases as predicted by the sea level history (Fig. 6-2) is rarely able to be deciphered. Within each cycle, the amount of induration is greatest just beneath the disconformity surface, where primary and secondary pore space may be almost totally occluded; this induration decreases markedly downward. Also, younger cycles show little cementation overall, but induration increases into lower, older cycles, reaching a maximum a few cycles (at 10 to 15 m subsurface depth) below the present platform top, where rocks are moderately well cemented and large vugs and channels are pervasive. The amount of leaching of aragonite and formation of secondary porosity appears to offset the amount of precipitated calcite cement resulting in little net loss of pore space (Beach, 1982, Harrison et al., 1984). The porosity of the nonskeletal (originally aragonitic) sediments averages 35%, whereas porosity of the skeletal (initially calcitic) sediments at depth averages 10 to 15%. Primary porosity passes downward into secondary porosity below about 10 m below the platform top (cf. Halley and Schmoker, 1983 for Florida). In the Bahamian cores, two and perhaps 3 stages of cementation tied to Quaternary sea level falls were recognized by Beach (1982). The first cements in cycles are recrystallized marine cements. Sea level fall causes deposition of meteoric, intergranular sparry cements and overgrowths, with inclusion-rich rinds on cements near the disconformity, and common rhizocretions. The cements subsequently may be overlain by infiltrated marine sediment following deposition of the next marine unit. Karstic vugs and channels lined with second generation cements are ascribed to the resultant sea level fall. With the next sea level rise and fall, a third generation of meteoric cements may develop (Beach, 1982). Much of the cementation could be vadose even though diagenesis is more rapid in the phreatic zone (Matthews and Frohlich, 1987; Budd, 1988), but vadose calcites (except for caliche) can rarely be distinguished from the phreatic calcites (Beach, 1982; Harrison et al., 1984), although vugs and channels lined with spar and silts likely are vadose.

In some beds, leaching and calcite cementation was followed by mixing-zone(?) dolomitization, while in younger units this pre-dolomite leaching phase is absent. Even with conservative dolomite production rates, modelling suggests that substantial amounts of mixing zone dolomites in these platforms may occur below subsurface depths of about 50 m to 100 m (Matthews and Frohlich, 1987; Humphrey and Quinn, 1989). It is unclear whether the bulk of dolomites in these platforms is mixing-zone (Ward and Halley, 1985; Humphrey, 1988), or related to marine water circulation through the platform (Whitaker and Smart, 1990 and this volume), or due to reflux of brines generated by evaporation during more arid phases (Berner, 1965; Goldstein et al., 1991b) or due to thermal convection of cool, deeper marine waters (Saller, 1984).

Example 3: Pennsylvanian Reservoirs of the U.S. Mid-continent:

Midcontinent cyclothems are a few meters to over 10 m thick, and formed under high amplitude ecstasy on a huge regional ramp, and are regionally traceable over much of the midcontinental U.S. (Figs. 6-4 to 6-7). The typical cyclothem consists of a basal regional disconformity, overlain by nearshore and deltaic clastics and transgressive-limestone, then black shales that formed in 60 to 100 m water depths; these shallow up into shallow water limestone bank/shoal or shallow shelf facies and may be capped by disconformities or by prograding clastics and a regional disconformity/paleosol (Fig. 6-4). In detail, many cyclothems contain smaller scale cycles capped by paleosols, which probably relate to smaller scale eustatic fluctuations superimposed on the larger 4th order sea level changes (Fig. 6-4 and 6-5). Modelling and mapping shows that many of these smaller cycles merge downdip with the major cyclothems (compare Fig. 6-4 and 6-7).

Heckel (1983) described a diagenetic model for U.S. Mid-continent cyclothems of Mid-Pennsylvanian age (Fig. 6-4). The diagenetic signature has characteristics of humid conditions with widespread sparry cements, and abundant leaching, to semi-arid conditions with caliche, local evaporites.

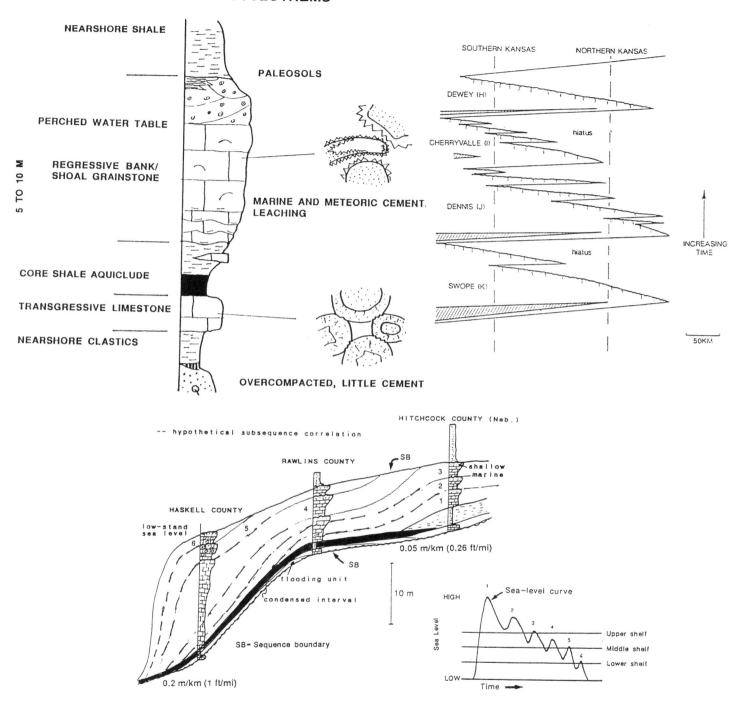

Fig. 6-4. Pennsylvanian cyclothems of the mid-continent. Top left: Schematic cycle of Pennsylvanian cyclothem showing vertical stacking of facies, and diagenetic overprint. After Heckel, 1983. Top right: Time-distance cross section showing relative sea level inundation associated with major cycles of the Kansas City Group, western Kansas. Condensed sections shaded; wavy lines at tops of packages are subaerially exposed bounding surfaces. (Watney et al., 1991).
Bottom: Conceptual model of Upper Pennsylvanian (Missourian) carbonate shelf cycle from ramp in western Kansas. Flooding unit (thin carbonate), condensed section (black shale) and shallowing upward (highstand) carbonates typify major cycles; minor cycles punctuate the major cycles (from Watney et al., 1991).

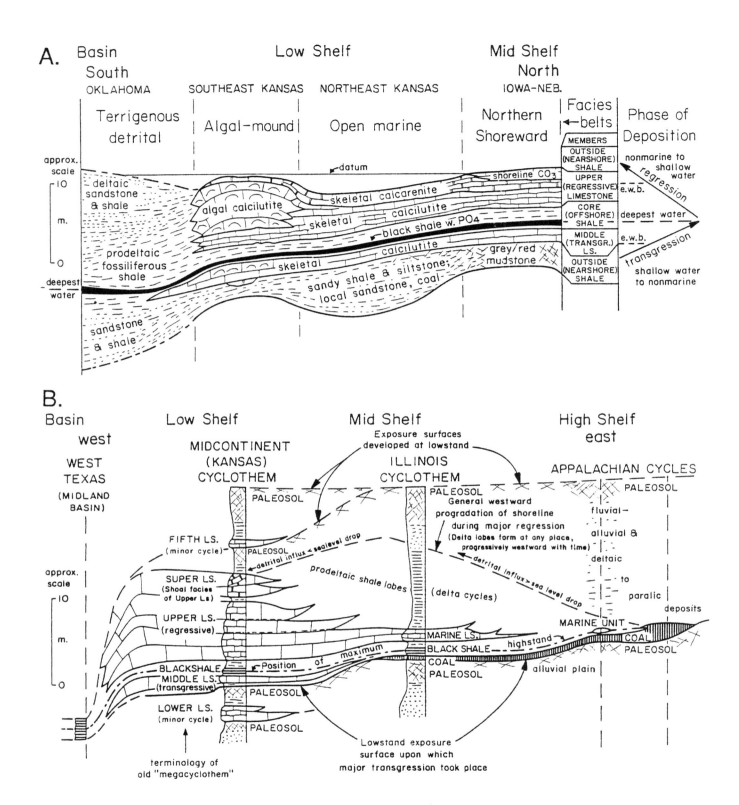

Fig. 6-5. A. Schematic regional cross-section of low- to mid shelf Upper Pennsylvanian cyclothem, U.S. Midcontinent. B. Schematic regional cross section of Pennsylvanian cyclothem from Texas through Midcontinent and into the Appalachians. Note the several minor paleosol horizons and the major cyclothem bounding paleosol (sequence boundary). From Heckel (1994).

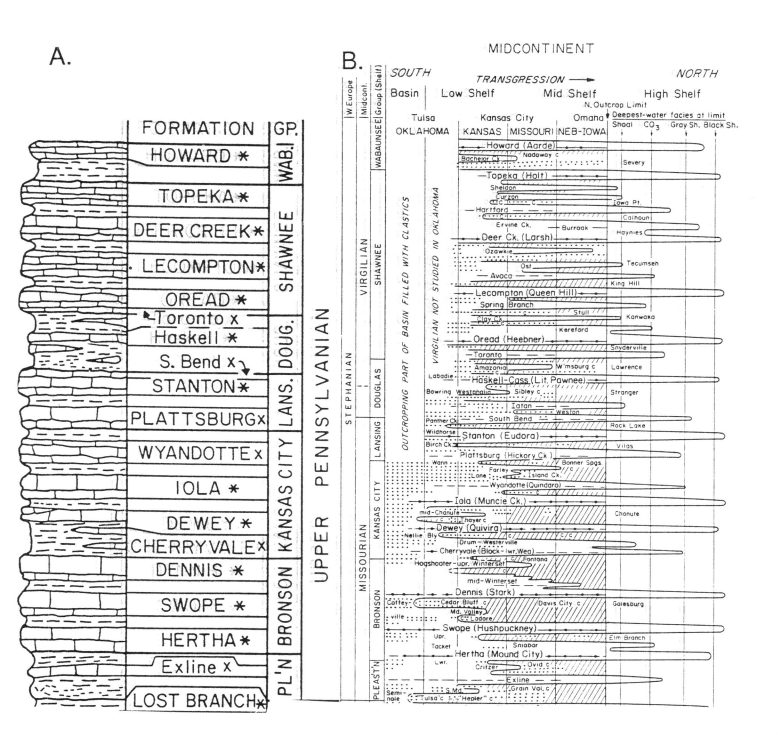

Fig. 6-6. Upper Pennsylvanian succession in Midcontinent, North America (left hand column) and the inferred sea level curve (right hand side). Subaerial exposure/paleosols - oblique line pattern; gray offshore shales - dashed horizontal lines; black offshore shales - solid horizontal lines; nonskeletal phosphorite - bold dots; terrigeneous detrital facies - fine dot pattern. From Heckel (1994).

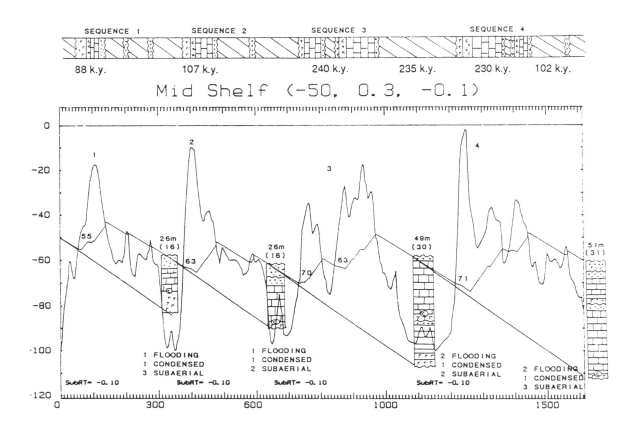
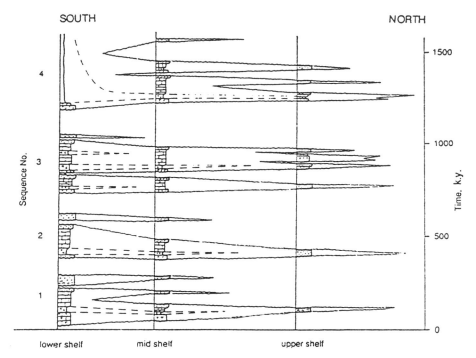

Fig. 6-7. Top: Computer simulation for a mid-shelf setting for Pennsylvanian (Missourian) cycles, western Kansas. Initial sediment elevation is -50 m, sedimentation rate is 0.3 m / k.y., subsidence rate is 0.1 m / k.y.
Bottom: Time-stratigraphic diagram for upper, middle and lower shelf settings generated by computer model. Note similarity to time-distance cross-section shown previously. Watney et al. (1991).

Cycles are a few meters to over ten meters thick and consist of basal near-shore shales, transgressive limestone, a core black shale (marking maximum marine flooding), regressive, coarsening upward, open marine limestones, bank- and shoal water facies, and a caliche or paleosol mudstone cap (Heckel, 1980; Watney et al., 1989).

Caliches, red paleosols, rhizoliths and leach fabrics were developed on cycles that formed under more arid paleoclimates in western Kansas, whereas karstification was more prevalent in the wetter paleoclimates of southeastern Kansas (Heckel, 1980; Watney et al., 1989, Goldstein et al., 1991a). The transgressive limestones below the core black shales show little early diagenesis and retained their aragonite mineralogy into early burial when they underwent pervasive overcompaction and neomorphism of aragonite to calcite (Fig. 6-4). The core shale apparently acted as an aquiclude, shielding the underlying transgressive limestone from downward penetration of meteoric water. In contrast, the upper, regressive limestones show much early marine cement, paleocaliche and soil formation, abundant leaching of aragonitic grains and pervasive blocky calcite cementation prior to much compaction (Fig. 6-4). This moldic and vuggy porosity is important in the Pennsylvanian reservoirs whereas calcite cementation and mud infiltration has occluded porosity in non-reservoir rocks (Ebanks and Watney, 1985). Similar leached reservoirs due to large scale migration of fresh-water lenses during high amplitude 4th order glacio-eustasy occur in regressive bryozoan and foram limestone portions of cyclothems above deeper water clayey facies in the Mid-Pennsylvanian Goen Limestone, Central Texas (Marquis and Laury, 1989). These facies have better porosity than the associated algal mound facies and early phreatic calcite cementation locally acted to partly occlude porosity (Marquis and Laury, 1989).

Example 4: Pennsylvanian Horseshoe Atoll Reservoirs, Texas:

The raised rim buildups on the southern and eastern margin of the Horseshoe Atoll (Fig. 6-8) are host to billion barrel fields that developed as a result of subsidence coupled with large, high frequency sea level fluctuations and attendant fresh water diagenesis (Schatzinger, 1983; Reid and Tomlinson-Reid, 1991). The buildups contain numerous erosionally bounded sequences (Fig. 6-8). Tidal flat, fenestral dolomitic mudstone were restricted to local shoals in the interior of the buildups on the elevated rim. Oolitic grainstones developed belt-like shoals over the buildups along the seaward margin of the rim and are major reservoir facies (Fig. 6-9). The buildup facies are high relief phylloid algal wackestone mounds bordered downslope by sponge-algal-bryozoan mounds (in water depths up to 20 m), which passed downslope into deeper shelf and slope muds. These muds form seals within and lateral to the buildups.

During the repeated sea level falls, the buildups and the facies distributions were erosionally modified (Fig. 6-9), islands developed weathered zones, paleosols and eolianites, and meteoric lenses caused leaching to form oomoldic and skelmoldic porosity, some dissolution of calcite cement, recrystallization of mud and grains to microcrystalline calcite, and secondary vug- and small cave development. Repeated exposure of the mounds even occurred during final drowning of the complex, attesting to the large fluctuations in sea level. The leached oolite grainstones have up to 30% porosity and up to 100 md permeability. Porosity and permeability is slightly lower in skeletal grainstones and low in the leached mound facies, even though these may have high porosity.

Example 5: Arid Zone Middle Pennsylvanian Reservoirs, Paradox Basin Cycles, U.S.A.:

Middle Pennsylvanian shelf cycles (Fig. 6-10 to 6-13) in the Paradox Basin form major reservoirs (Choquette and Traut, 1963; Pray and Wray, 1963; Wilson, 1975; Weber, Sarg and Wright, this volume). The shelf carbonates contain large scale, regionally correlative 4th order cycles up to thirty m thick (Fig. 6-10) and possibly 100 to 400 k.y. duration, that contain higher frequency

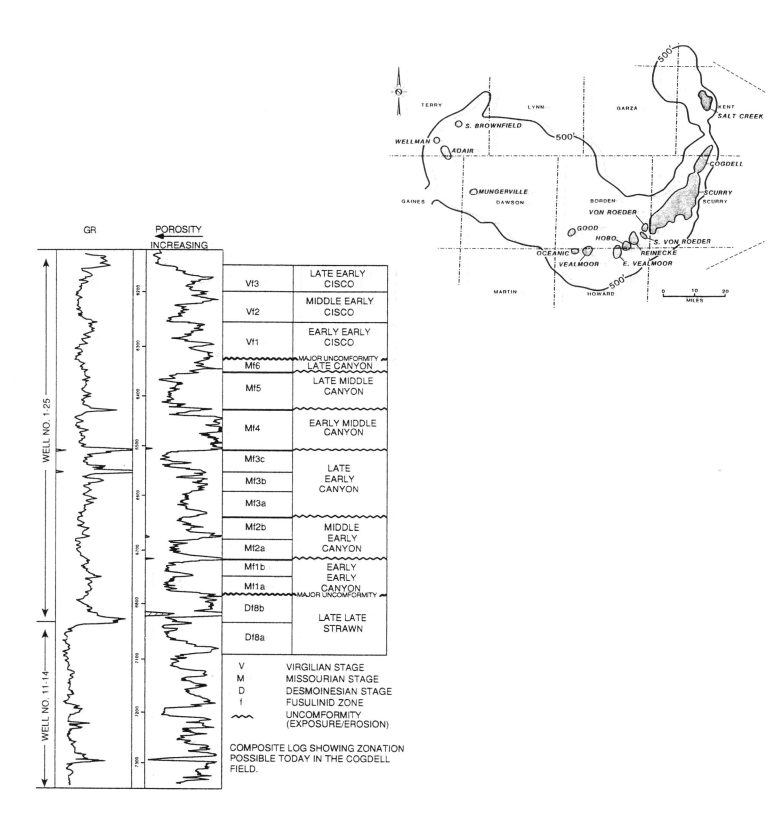

Fig. 6-8. Top: Location of Cogdell Field, Horseshoe Atoll, Texas. This raised rim around the atoll has a local relief exceeding 1000 to 1500 ft. above the regional platform Bottom: Composite gamma ray and porosity logs of the Cogdell Field, and position of unconformities which punctuate the section.

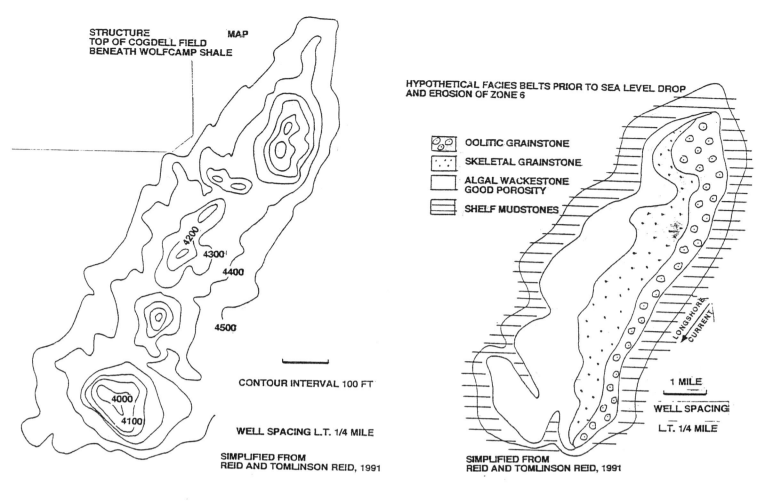
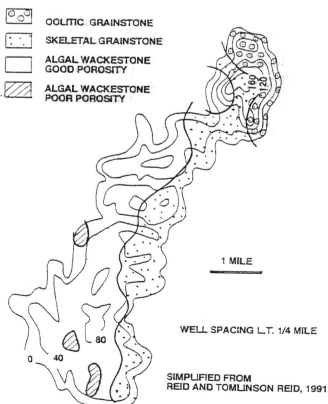

Fig. 6-9. Top left: Structure and subcrop map of the top of the Cogdell Field beneath the Wolfcamp Shale. Top right: hypothetical facies distribution Middle to Early Canyon (Zone 6) interval prior to sea level fall and erosion. Bottom: Present day isopach and facies map of Middle to Early Canyon interval, showing facies distribution after sea level fall and erosional modification. From Reid and Tomlinson-Reid, 1991.

(ave. 6m thick) cycles possibly 20 to 40 k.y. duration (Goldhammer et al., 1991). Cycle facies (Fig. 6-10 and 6-11) include basal reworked sandstone, deep water black laminated dolomitic shale and argillaceous mudstone, spiculitic muds with packstone lenses, and phylloid algal, mud-rich mounds 6 to 12 m thick, with non-skeletal and skeletal packstone/grainstone caps. Mounding is common in the cycles, and is typical of high amplitude eustasy and reflects rapid sea level rises which generates much accommodation over the shelf. Antecedent tectonic or depositional highs localize many buildups. Most cycle-boundaries are marked by subaerial exposure surfaces or in downslope positions, by low-stand algal laminites, evaporites and clastics. Caliches are developed on cycles associated with regionally correlatable 4th order sequence boundaries and the regional black laminated mudstones are associated with major 4th order deepening events (Fig. 6-10 to 6-13). Basin sections contain 29 correlative shale-evaporite 4th order cycles.

Upper parts of cycles commonly form the reservoirs. Meteoric diagenesis caused by major sea level falls is evident in upper parts of shelf cycles, which show much leaching of aragonite in algal bank facies, skeletal packstone/wackestone and non-skeletal grainstone/packstone caps (Wilson, 1975), and much non-luminescent, non-ferroan sparry cement in Ismay reservoirs described by Dawson (1988). The restriction of the meteoric diagenesis to upper parts of cycles even though 4th order eustatic sea level falls of 50 to 60 m were involved, probably relates to gradual subsidence diminishing the relative sea level fall to 20 m or so (Goldhammer, et al., 1991). Shaly and muddy carbonates low in the cycles probably also acted as aquitards, preventing deep penetration of meteoric waters. The evidence for meteoric diagenesis in these cycles is interesting in view of the arid conditions of deposition of time-equivalent evaporites in the Paradox Basin.

In contrast, Middle Pennsylvanian subsurface algal mounds of the Bug and Papoose Canyon fields of the Paradox basin, Utah and Colorado show evidence of more typical arid zone diagenesis (Roylance, 1990). The mounds underwent marine aragonite cementation, followed by some leaching of capping facies above mounds, brecciation and dolomitization; there was little sparry calcite cementation, in spite of substantial sea level fluctuations, and there was considerable porosity reduction by anhydrite and less common halite cements.

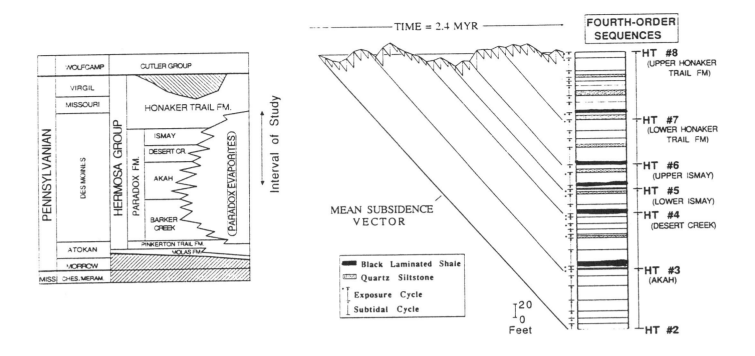

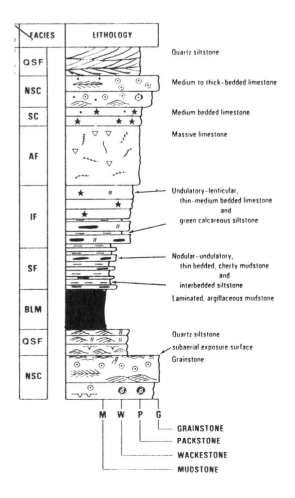

Fig.6-10. Pennsylvanian Paradox Basin cycles (from Goldhammer et al., 1991): Top left: Pennsylvanian chronostratigraphic diagram for the Paradox Basin. Top right: simplified stratigraphy of the Pennsylvanian Honaker Trail section, and its Fischer plot; straight line connecting base of Fischer plot to base of section is mean subsidence vector. Bottom: Idealized 5th order cycle, Paradox Basin.

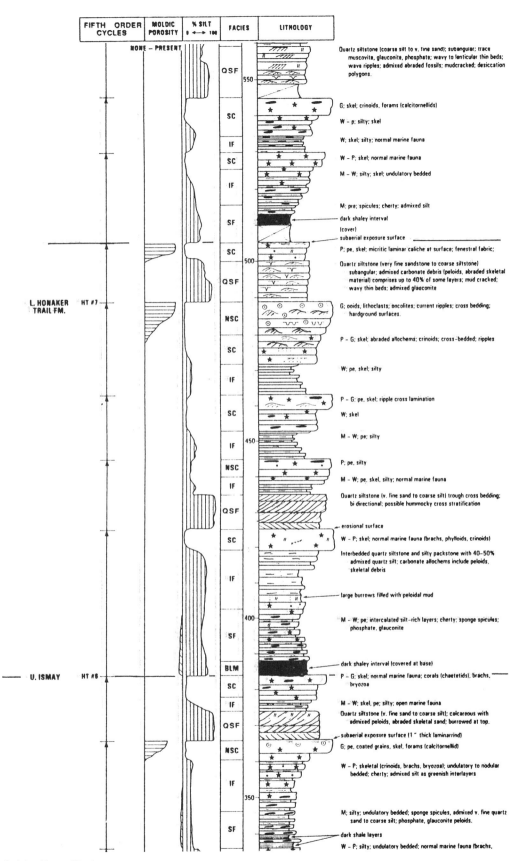

Fig. 6-11. Detailed measured section of Pennsylvanian Paradox Basin L. Honaker Trail Formation, which represents a 4th order cycle bounded by deeper water laminated argillaceous lime mudstone (black). Smaller scale cycles shown alongside column by vertical arrows. From Goldhammer et al. (1991).

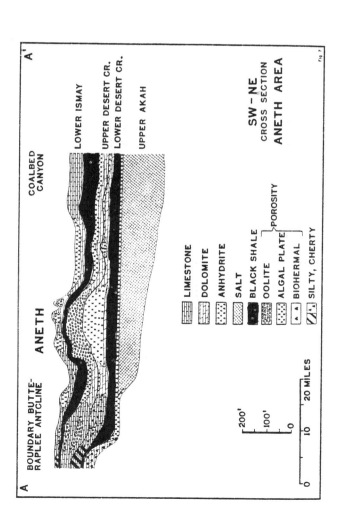

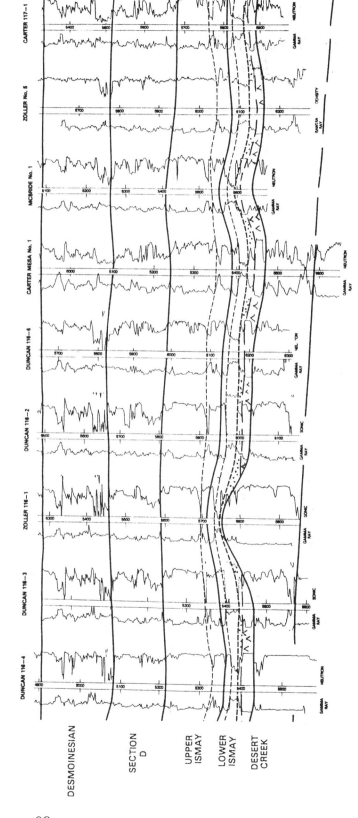

Fig.6-12. Left: Schematic cross-section of Pennsylvanian Ismay, Desert Creek and Upper Akah cycles showing relation between shelf mound development and evaporites. Paradox Basin. From Peterson and Ohlen, 1964. Below: Part of regional cross section across McElmo Creek (Aneth Field), showing 4th order sequence boundaries (solid lines), 4th order TST tops (thin dashed lines); in Lower Ismay, LST contains evaporites (inverted V's) and sandstone (dots). The Upper Ismay also contains a similar LST but the detailed lithologies are not shown. Scale bars are 100 ft intervals. These 4th order cycles are made up of small 5th order cycles that progressively shallow upward. (from Goldhammer et al., 1991).

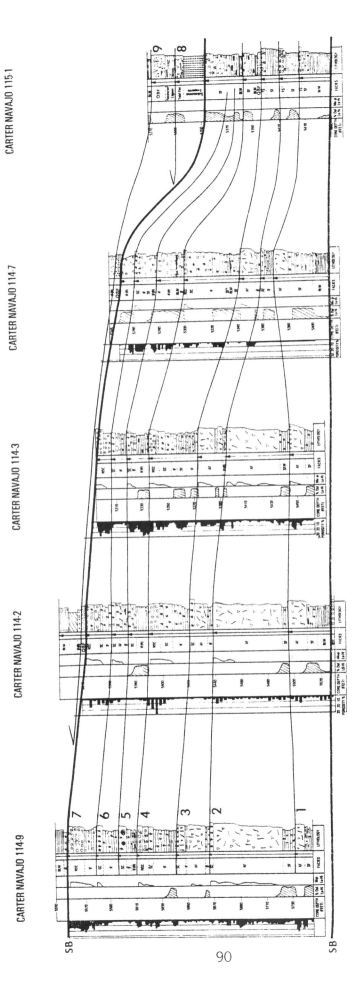

Fig. 6-13. Part of shelf-to-basin cross-section of the Desert Creek interval. Thick lines are 4th order sequence boundaries; smaller scale shallowing upward cycles shown by thin lines. From Goldhammer et al., 1991

CHAPTER 7

RESERVOIRS AND DIAGENESIS ASSOCIATED WITH MAJOR SEQUENCE-BOUNDING UNCONFORMITIES

This chapter briefly reviews how carbonates are modified beneath regional unconformities, using two examples. Regional uplift or long term sea level falls form widespread regional unconformities. Given sufficient rainfall, the unconformity surface acts as a recharge area for regional aquifers, which strongly controls early diagenesis in the sequence (Fig. 7-1). In newly deposited sediments, the water movement is via intergranular pores by diffuse flow. However, with increasing time and increasing cementation and leaching, water movement becomes dominated by conduit flow along widening joints and caves undergoing dissolution beneath the karst surface. Pleistocene mature karstic terrains include those documented by Purdy (1974a, b), and Smart et al. (1988), from British Honduras and there are many examples in the humid climates of SE Asia (Horbury, pers. obs.). Detailed discussions of geological aspects of paleokarst are given in James and Choquette (1988) and Wright, Estaban, and Smart (1991).

Key Features

1. Diagenesis related to major unconformities can extend tens to some hundreds of meters downsection.

2. In newly emergent terrains, intergranular porosity in diffuse flow aquifers can be plugged by early meteoric phreatic cements, whose abundance decreases downsection and downflow. However, development of chalky and moldic porosity may offset this porosity decrease.

3. Beneath unconformities that are emergent for millions of years, karstic relief, breccia-filled cave systems with both marine and non-marine fills, and burial induced fracturing of cave-roofs may form extremely compartmentalized reservoirs.

4. In the subsurface in non-cored well data, it is easier to identify the products of karstification such as porosity development/occlusion, rather than the surface itself (Estaban and Klappa, 1983). Where clastic or basinal facies overlie the unconformity, a high gamma log count due to glauconite or uranium which may mark the flooding surface. The unconformity may be evident in a marked biostratigraphic break. In seismic reflection profiles, unconformity relief will be difficult to distinguish from relief associated with pinnacle reefs or other high relief buildups.

<u>Example 1: Unconformity-Related Diagenesis Associated with Diffuse Flow Aquifers, Late Mississippian Reservoirs of the Appalachians:</u>

Late Mississippian carbonates of the Appalachians were cemented in diffuse flow aquifers during 2 major phases of unconformity development (Niemann and Read, 1988; Nelson and Read, 1989). These ramp carbonates contain regionally traceable shallow burial calcite cements that are early non-ferroan, CL-banded zones extending down into the carbonates from a few meters to over 60 m (Fig. 7-2). The cements can be traced over thousands of square kms and from tens to over 200 kms downdip. The cement zones are 1. nonluminescent 2. luminescent and 3. nonluminescent cement. These "zones" commonly contain thin CL-defined cement laminae which are not regionally correlative. In the Appalachians, the nonluminescent cements pass downdip into dull luminescent cements (Fig. 7-2). The cement stratigraphy can still be recognized in these dull CL cements using Dickson's stains sensitive to Fe content of the calcites (Nelson and Read, 1989). The nonluminescent-dull-nonluminescent zonation of updip areas is equivalent to pink-purple-pink staining, dull CL zones downdip. In the Appalachians, the early

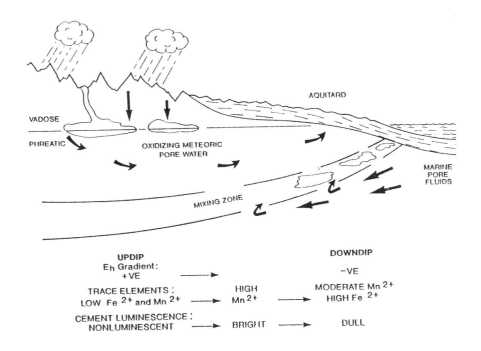

Fig. 7-1. Schematic diagram of a carbonate aquifer that is unconfined updip and confined by aquitard downdip. The regional karstic unconformity acts as a recharge area. Numerous caves extend down to and are localized by groundwater table, and by the coastal mixing zone. Regional gradients in Eh, reduced Fe and Mn, and cathodoluminescence of clear calcite cements are shown.

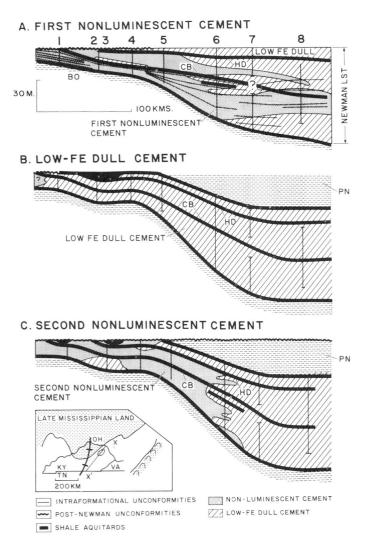

Fig. 7-2. Schematic model showing the depositional history of meteoric cements in the diffuse flow paleoaquifer in the Mississippian Newman Limestone, Kentucky (Niemann and Read, 1988). The first generation of cements (A) formed from an aquifer sourced from above the Glen Dean Limestone in the Late Mississippian; this aquifer had oxidizing waters updip (shaded areas) and more reducing waters downdip (cross hatched areas). Following Late Mississippian Pennington transgression, this aquifer shifted updip, resulting in more reducing waters in the carbonates, and Fe rich cements (B). With the Mississippian-Pennsylvanian unconformity development, a major regional aquifer generated the next generation of cements from oxidizing groundwaters that passed downdip into more reducing waters.

nonferroan zoned cements fill up to 80% of the pore space updip, decreasing to less than 40%, 80 kms to the south away from the postulated recharge area.

The nonluminescent-luminescent-nonluminescent cement zones all are low in iron, and could have formed from oxidizing pore waters sourced from the Late Mississippian and Mississippian-Pennsylvanian unconformities capping the sections (Niemann and Read, 1988). At this time the climate in the Appalachians was becoming wet and tropical, and deltas were prograding out over the carbonate ramp. The lower of these two unconformities sourced the waters for the zone 1 nonluminescent cements. The succeeding nonferroan luminescent cement formed during the subsequent Late Mississippian transgression which caused stagnation of the aquifer. The later nonluminescent zone formed during development of the Mississippian-Pennsylvanian unconformity. The updip nonluminescent cements (zones 1 and 3) are interpreted to have time-equivalent dull CL cements downdip where waters were more reducing. Although there are unconformities lower in the Mississippian carbonates in the Appalachians, these had little effect on cementation because they developed during arid conditions (Niemann and Read, 1988). These unconformity-sourced cements tended to plug primary porosity updip, but appear to have caused only limited plugging of pore space downdip where much of the Mississippian produces. The aquifer waters also caused metastable mineralogies to be converted to low Mg calcite updip, and could have formed chalky porosity in some of the updip ooid shoals in the Appalachians.

Note that not all shallow burial calcites are deposited from meteoric waters sourced from unconformities, but may in some cases be sourced from refluxing brines associated with evaporites higher in the section (Goldstein et al., 1991) or from waters sourced from tectonic highlands peripheral to foreland basins, with flow via aquifers (karstic zones, quartz sands, or carbonate sands) (Grover and Read, 1983; Dorobek, 1987).

Example 2: Reservoirs Associated with Mature Karstic Terrains, Ordovician Knox-Ellenburger Unconformity, U.S.A.:

The Knox-Ellenburger unconformity that caps the Cambro-Ordovician sequence throughout much of North America (Mussman et al., 1988; Kerans, 1988; Knight, James and Lane, 1991) has localized important hydrocarbon and Pb-Zn deposits. It is up to 10 m.y. duration, and is the boundary to the "Sauk Sequence" of Sloss (1963), one of the fundamental 2nd order cycles of North America (Vail et al., 1977). The unconformity formed during a global sea level fall and in the Appalachians, during uplift associated with a subduction-related peripheral bulge. The unconformity developed during a time of increasingly humid climate, in contrast to the semi-arid conditions that typified Knox cyclic deposition. The unconformity surface locally has over 130 m of erosional relief. There are well developed cavern- and -sinkhole fills (both subaerial and margin, and intraformational breccias due to dissolution of limestone and collapse of the overlying dolomite (Mussman et al., 1988). Extensive collapse breccias tens to 250 m beneath the unconformity formed in the meteoric phreatic and mixing zone associated with a thick zone (up to 100 m) of meteoric ground water.

Nonluminescent cements occur down to 200 m below the unconformity surface (Mussman et al., 1988). Down section, the nonluminescent cements are absent, and the limestones are cemented by dully luminescent cement. These cements tended to plug porosity in the upper 200 m of section, so that the main porosity was karst-related vug, fissure and cavernous porosity. The initial non-luminescent cement sequence probably relates to development of a paleoaquifer whose waters were oxidizing to depths of 100 to 200 m below the unconformity reflecting the humid setting and the highly permeable character of the carbonates. The meandering networks of caves and intraformational breccias probably carried large volumes of meteoric water into the subsurface. With stagnation of the aquifer, possibly during Middle Ordovician transgression, bright cements were deposited as the waters became more reducing. A second non-luminescent cement in the upper Knox carbonates was due to regeneration of the paleoaquifer system by upland sourced meteoric waters which caused cementation in the overlying Middle Ordovician carbonates (Grover and Read, 1983).

Paleokarstic reservoirs occur in the Early Ordovician Ellenburger Group, Texas. These reservoirs have cumulative production of 1.4 billion barrels of oil through 1985 with low recovery efficiencies due to extreme horizontal and vertical compartmentalization of reservoirs. They have been described in detail by Kerans (1988). Most reservoirs occur in the upper 60 to 150 m of the unit regardless of the original peritidal depositional facies and occupy laterally extensive breccia zones (Fig. 7-3A). They contain a lower collapse zone of chaotic dolostone clast-support breccias of highly variable thickness, averaging 15 to 30 m, with zones of unbrecciated dolomite in the lower part (Figs. 7-3A,B). Porosity is from 1 to 15%. These are overlain by a middle zone of cave fill, siliciclastic-matrix supported chaotic breccias of Ellenburger dolomite and clasts of shale and sandstone (from the overlying Simpson Group), and intervals of crudely bedded sandstone and shale with local soft-sediment folds and faults (Figs. 7-3A,B). This siliciclastic middle zone has a distinctive log signature and also forms a low permeability barrier. The upper zone, the cave roof, is highly fractured dolomite with fitted- to rotated clast fabrics. Porosity is from 2 to 20% and the zone is the major pay interval.

The karstic cave systems developed at a stable paleogroundwater table 30 to 60 m below the top of the Ellenburger. The lower breccias are cave-collapse, stope-deposits whereas the middle matrix rich zone is a Middle Ordovician marine cave fill, analogous to present day blue holes. The cave-roof fracture breccias were formed during later burial compaction and fracturing of the cave roof. The breccia distribution and muddy cave fills cause the high degree of compartmentalization of these reservoirs.

ACKNOWLEDGMENTS

I would like to thank Anna Balog, Aus Al-Tawil, Mike Pope, L. B. Smith, Art Hall and Ellen Mathena for help in preparing this paper. I would like to thank my present and former students for many of the ideas and discussions outlined here.

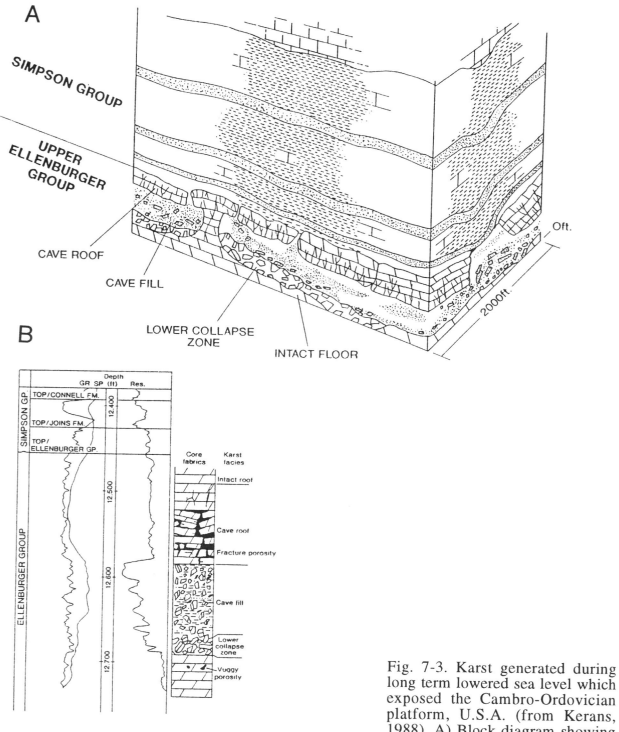

Fig. 7-3. Karst generated during long term lowered sea level which exposed the Cambro-Ordovician platform, U.S.A. (from Kerans, 1988). A) Block diagram showing sub-unconformity cave system. B) Comparison of log signature and karst facies in Gulf 000-1 TXL well, Emma Ellenburger reservoir. Much of the porosity is in the fractured cave roof.

BIBLIOGRAPHY

Balog, A., Haas, J., Read, J.F., and Coruh, C., in press, Shallow marine record of orbitally forced cyclicity in a Late Triassic carbonate platform, Hungary: J. Sed. Research.

Beach, D. K., 1982, Depositional and diagenetic history of Pliocene-Pleistocene carbonates of northwestern Great Bahama Bank; Evolution of a carbonate platform: Ph.D. Dissertation, University of Miami, Fischer Island Station, 447 pp.

Berger, A., Loutre, M. F., and Dehant, V., 1989, Influence of the changing lunar orbit on the astronomical frequencies of pre Quaternary insolation patterns: Paleoceanography, v. 4, p. 555-564

Berner, R. A., 1965, Dolomitization of Mid-Pacific atolls: Science, v. 147, p. 1297-1299.

Berner, R. A., 1991, A model for atmospheric CO_2 over Phanerozoic time: American Journal of Science, v. 291, p. 339-376.

Borer, J. M. and Harris, P. M., 1991, Lithofacies and cyclicity of the Yates Formation, Permian Basin: Implications for reservoir heterogeneity: A.A.P.G. Bull. v. 75, p. 726-779.

Bond, G. C., Kominz, M.A., and Beavan, J., 1991, Evidence for orbital forcing of Middle Cambrian peritidal cycles: Wah Wah range, south-central Utah: in E.K. Franseen, W.L. Watney, C.G.St.C. Kendall, and W.Ross, eds., Sedimentary Modelling, Kansas Geol. Surv. Bull., v. 233, p. 293-318.

Bubb, J. N. and Hatfield, W. G., 1977, Seismic Recognition of Carbonate Buildups. In Seismic Stratigraphy - Applications to Hydrocarbon Exploration, Ed. C.E. Payton, AAPG Mem. 26, p. 185-204.

Budd, D.A., 1988, Aragonite-to-calcite transformation during fresh-water diagenesis of carbonates: insights from pore-water chemistry: Geol. Soc. America Bull., v. 100, p. 1260-1270.

Budd, D. A. and Land, L. S., 1990, Geochemical imprint of meteoric diagenesis in Holocene ooid sands, Schooner Cays, Bahamas: Correlation of calcite cement geochemistry with extant groundwaters: Jour. Sedimentary Petrology, v. 60, p. 361-378.

Burchette, T. P. and Wright, V. P., 1992, Carbonate ramp depositional systems: Sedimentary Geology, v. 79, p. 3-57.

Choquette, P. W. and Traut, J. D., 1963, Pennsylvanian carbonate reservoirs, Ismay field, Utah and Colorado, in R.O. Bass, ed., Shelf Carbonates of the Paradox Basin, Four Corners Geological Society, 4th Field Conference Gdbk., p. 157-184.

Chuber, S. and Pusey, W. C., 1985, Productive Permian carbonate cycles, San Andres Formation, Reeves Field, West Texas: in P.O. Roehl and P.W. Choquette, Carbonate Petroleum Reservoirs, Springer-Verlag, New York, p. 289-308.

Collins, L. B., 1988, Sediments and history of the Rottnest Shelf, southwest Australia: a swell-dominated non-tropical carbonate margin: Sedimentary Geology, v. 60, p. 15-49.

Crevello, P. D., 1991, High frequency carbonate cycles and stacking patterns: Interplay of orbital forcing and subsidence on Lower Jurassic rift platforms, High Altals, Morocco, In E.K. Franseen, W.L. Watney, C.G.St.C.Kendall, and W.H. Ross, Sedimentary Modelling: Kansas Geol. Surv. Bull. 233, p. 207-230.

Crowley, T. J. and Baum, S. K., 1991, Toward reconciliation of Late Ordovician glaciation with very high CO_2 levels: Journal of Geophysical Research, v. 96, p. 22,597-22,610.

Davies, J. R., 1991, Karstification and pedogenesis on a Late Dinantian carbonate platform, Anglesey, North Wales: Proc. Yorkshire Geol. Soc., v. 48, p. 297-322.

Dawson, W. C., 1988, Ismay reservoirs, Paradox Basin - diagenesis and porosity development: Rocky Mountain Association of Geologists, 1988 Carbonate Symposium, p. 163-174.

Demicco, R. V., 1985, Patterns of platform and off-platform carbonates of the Upper Cambrian of western Maryland: Sedimentology, v. 32, p. 1-22.

Dill, R. F., 1977, Blue holes - Geologically significant submerged sinkholes and caves off British Honduras and Andros, Bahama Islands, In: Proceedings, Third International Coral Reef Symposium, Rosenstiel School of Marine and Atmospheric Science, University of Miami, Florida, p. 237-242.

Dorobek, S. L., 1987, Petrography, geochemistry and origin of burial diagenetic facies, Siluro-Devonian Helderberg Group (carbonate rocks), Central Appalachians (abs.): AAPG Bull., v. 71, p. 492-514.

Dorobek, S. L., Smith, T. M., and Whitsitt, P.M., 1993, Microfabrics and geochemical signatures associated with meteorically altered dolomite: examples from Devonian and Mississippian carbonates of Montana and Idaho: in R. Rezak and D. Lavoie, eds., Carbonate Microfabrics: Frontiers in Sedimentary Geology Series, Springer-Verlag, New York, p. 205-225.

Dorobek, S. L., 1995, Synorogenic carbonate platforms and reefs in foreland basins: Controls on stratigraphic evolution and platform morphology: in Stratigraphic Evolution of Foreland Basins, SEPM Spec. Publ. No. 52, p. 127-147.

Dunham, R. J., 1969, Early vadose silt in Townsend Mound (Reef), New Mexico, In: G.M. Friedman, ed., Depositional Environments in Carbonate Rocks, Soc. Econ. Paleontologists and Mineralogists Spec. Publ. No. 14, p. 139-181.

Ebanks, W. J., Jr. and Watney, W. L., 1985, Geology of Upper Pennsylvanian carbonate oil reservoirs, Happy and Seberger fields, northwestern Kansas: in P.R. Roehl and P.W. Choquette, eds., Carbonate Petroleum Reservoirs, Springer-Verlag, New York, p. 239-250.

Einsele, G. and Ricken, W., 1991, Limestone-marl alternation - an overview. In Cycles and Events in Stratigraphy (eds., Einsele, G., Ricken, W., and Seilacher, A.). Springer Verlag, New York, p. 23-47.

Elrick, M. E. and Read, J. F., 1991, Development of cyclic ramp-to-basin carbonate deposits, Lower Mississippian, Wyoming and Montana: unpub. Ph.D. dissertation, Virginia Polytechnic Institute and State University, Blacksburg, Virginia, 169 p.

Elrick, M. E., Read, J. F., and Çoruh, C., 1990, Short-term paleoclimatic fluctuations expressed in Lower Mississippian ramp-slope deposits, southwestern Montana: Geology, v. 19, p. 799-802.

Estaban, M. and Klappa, C. F., 1983, Subaerial exposure environment: in P.A. Scholle, D.G. Bebout and C.H.

Moore, Carbonate Depositional Environments, AAPG Memoir 33, p. 1-54.

Ettensohn, F. R., Dever, G. R. Jr., and Grow, J. S., A paleosol interpretation for profiles exhibiting subaerial exposure 'crusts' from the Mississippian of the Appalachian Basin: in Juergen Reinhardt and Wayne R. Sigleo, eds., Paleosols and Weathering through Geologic Time, Geol. Soc. America Spec. Paper 216, p. 49-80.

Fischer, A. G., 1964, The Lofer cyclothems of the Alpine Triassic: Kansas State Geological Surv. Bull. 169, v. 1, p. 107-150.

Fischer, A. G., 1982, Long-term climatic oscillations recorded in stratigraphy, in Climate in Earth History, National Academy Press, Washington D.C., p. 97-104.

Frakes, L. A. and Francis, J. E., 1988, A guide to Phanerozoic cold polar climates from high-latitude ice-rafting in the Cretaceous: Nature, v. 333, p. 547-549.

Friedman, G. M., 1980, Dolomite is an evaporative mineral: Evidence from the rock record and from sea-marginal ponds of the Red Sea: in D.H. Zenger, J.B. Dunham and R.L. Ethington, eds., Concepts of Dolomitization: SEPM Spec. Publ. 28, p. 69-80.

Ginsburg, R. N., 1971, Landward movement of carbonate mud: new model for regressive cycles in carbonates (abst.): AAPG Bull., v. 55, p. 340.

Ginsburg, R. N., Swart, P. K., Eberli, G. P., NcNeill, D. F., and Kenter, J.A.M., 1990, Bahamas Drilling Project Phase 1, Appendix 1 and 2, Comparative Sedimentology Laboratory, Rosenstiel School of Marine and Atmospheric Science, University of Miami, 32 p.

Goldhammer, R. K., Dunn, P. A., and Hardie, L. A., 1990, Depositional cycles, composite sea-level changes, cycles stacking patterns, and the hierarchy of stratigraphic forcing: Examples from Alpine Triassic carbonates: Geol. Soc. America Bulletin, v. 102, p. 535-562.

Goldhammer, R. K., Oswald, E. K., and Dunn, P.A., 1991, Hierarchy of stratigraphic forcing: Example from the Middle Pennsylvanian shelf carbonates of the Paradox Basin: in E.K. Franseen, W.L. Watney, C.G.St.C. Kendall, and W.Ross, eds., Sedimentary Modelling, Kansas Geol. Surv. Bull., v. 233, p. 361-414.

Goldhammer, R. K., Oswald, E. J., and Dunn, P. A., 1991, Hierarchy of stratigraphic forcing: Example from Middle Pennsylvanian shelf carbonates of the Paradox Basin: in E.K. Franseen et al., Sedimentary Modelling: Computer simulations and methods for improved parameter definition: Kansas Geological Survey Bull. 233, p. 361-413.

Goldhammer, R. K., Lehmann, P.J., and Dunn, P. A., 1993, The origin of high frequency platform carbonate cycles and third order sequences (Lower Ordovician El Paso Group, West Texas): Constraints from outcrop data and modelling: Jour. Sediment. Petrology, v. 63, p. 318-359.

Goldhammer, R.K., Harris, M.T., Dunn, P.A., and Hardie, L.A., 1993, Sequence stratigraphy and systems tract development of the Latemar Platform, Middle Triassic of the Dolomites (Northern Italy): Outcrop calibration keyed by cycle stacking patterns: in R.G. Loucks and J. Frederick Sarg, eds., Carbonate Sequence Stratigraphy, AAPG, p. 353-387.

Goldstein, R. H., Anderson, J. E., and Bowman, M. W., 1991a, Diagenetic responses to sea-level change: Integration of field, stable-isotope, paleosol, paleokarst, fluid inclusion, and cement stratigraphy research to determine history and magnitude of sea-level fluctuation: in E.K. Franseen et al., Sedimentary Modelling: computer simulations and methods for improved parameter definition: Kansas Geological Survey Bull. 233, p. 139-162.

Goldstein, R. H., Stephens, B. P., and Lehrmann, D. J., 1991b, Fluid inclusions elucidate conditions of dolomitization in Eocene of Enewetak atoll and mid-Cretaceous Valles Platform of Mexico (Abstract): in A. Bosellini et al., eds., Dolomieu Conference on Carbonate Platforms and Dolomitization, Ortisei/St. Ulrich, Val Gardena/Grodental, The Dolomites, Italy, p. 93.

Goodwin, P. W. and Anderson, E. J., 1985, Punctuated aggradational cycles: A general hypothesis of episodic stratigraphic accumulation: Journal of Geology, v. 93, p. 515-533.

Grover, G. A. Jr. and Read, J. F., 1983, Paleoaquifer and deep burial cements defined by cathodoluminescent patterns, Middle Ordovician carbonates, Virginia: AAPG Bull., v. 78, p. 1275-1303.

Halley, R. B. and Schmoker, J. W., 1983, High-porosity carbonate rocks of South Florida: progressive loss of porosity with depth: A.A.P.G. Bull. v. 67, p. 191-200.

Handford, C. R., 1988, Review of carbonate sand-belt deposition of ooid grainstones and application to Mississippian reservoirs, Damme Field, southwestern Kansas: AAPG Bull. v.72, p.1184-1199.

Handford, C. R. and Loucks, R. G., 1994, Carbonate depositional sequences and systems tracts: in Carbonate Sequence Stratigraphy, AAPG, p. 3-41.

Haq, B. U. et al., 1987, Mesozoic-Cenozoic cycle chart: in A.W. Bally, ed., Atlas of Seismic Stratigraphy, AAPG Studies in Geology #27. 124 p.

Hardie, L. A., Bosellini, A., and Goldhammer, R. K., 1986, Repeated subaerial exposure of subtidal carbonate platforms, Triassic, Northern Italy: evidence for high-frequency sea level oscillations on a 10 year scale, Paleoceanography, 1:447-457.

Hardie, L. A., Dunn, P. A., and Goldhammer, R.K., 1991, Field and modelling studies of Cambrian carbonate cycles, Virginia Appalachians - discussion: Journal of Sedimentary Petrology, v. 61, p. 636-646.

Harris, P. M., Flynn, P. E., and Sieverding, J. L., 1988, Mission Canyon (Mississippian) reservoir study, Whitney Canyon-Carter Creek field, southwestern Montana, in A.J. Lomando and P.M. Harris, Giant Oil and Gas Fields - A Çore Workshop: SEPM Core Workshop 12, v. 2, p. 695-740.

Harris, P. M. and Stoudt, E. L., 1988, Stratigraphy and lithofacies of the San Andres Formation, C.S.Dean "A", XIT, and SW Levelland units of Levelland-Slaughter field, Permian Basin: in A.J. Lomando and P.M. Harris, Giant Oil and Gas Fields, A Core Workshop, SEPM Core Workshop No. 12, p. 649-694.

Harrison, R. S., Cooper, L. D., and Coniglio, M., 1984, Late Pleistocene carbonates of the Florida Keys: in Carbonates in Subsurface and Outcrop, 1984 CSPG Core Conference, Canadian Society of Petroleum Geologists, Calgary, Alberta, Canada, p. 291-306.

Harrison, R. S. and Steinen, R. P., 1978, Subaerial crusts, caliche profiles and breccia horizons: comparison of some Holocene and Mississippian exposure surfaces, Barbados and Kentucky: Geol. Soc. America Bull., v. 89, p. 389-396.

Heckel, P. H., 1980, Paleogeography of eustatic model for deposition of midcontinent Upper Pennsylvanian cyclothems: in T.D. Fouch and E.R. Magathan, eds., Paleozoic Paleogeography of west-central United States, West-Central United States Paleogeography Symposium 1, Rocky Mountain Section SEPM, p. 197-215.

Heckel, P. H., 1983, Diagenetic model for carbonate rocks in Mid-continent Pennsylvanian eustatic cyclothems: Jour. Sedimentary Petrology, v. 53, p. 733-759.

Heckel, P., 1985, Recent interpretations of Late Paleozoic cyclothems: in Proceedings of the Third Annual Field Conference, Mid-Continent Section, SEPM, Lawrence, Kansas, p. 1-22.

Heckel, P. H., 1994, Evaluation of evidence for glacio-eustatic control over marine Pennsylvanian cyclothems in North America and consideration of possible tectonic effects: in Tectonic and Eustatic Controls on Sedimentary Cycles, SEPM Concepts in Sedimentology and Paleontology #4, Society for Sedimentary Geology, p. 65-87.

Horbury, A. D., 1987, Sedimentology of the Urswick Limestone Formation in South Cumbria and North Lancashire. Unpublished Ph.D. Thesis, University of Manchester, 668 p.

Horbury, A. D., 1989, The relative roles of tectonism and eustasy in the deposition of the Urswick Limestone Formation in South Cumbria and north Lancashire. In: R.S. Arthurton, P. Gutteridge and S.C. Nolan, eds., The Role of Tectonics in Devonian and Carboniferous Sedimentation in the British Isles. Yorkshire Geol. Soc. Occasional Publ. no. 6, p. 153-169.

Horbury, A. D. and Adams, A. E., 1989, Meteoric phreatic diagenesis in cyclic late Dinantian carbonates, northwest England: Sedimentary Geology, v. 65, p. 319-344.

Hovorka, S. D., Nance, H. S., and Kerans, C., 1993, Parasequence geometry as a control on permeability evolution: examples from the San Andres and Grayburg Formations in the Guadalupe Mountains, New Mexico: The University of Texas at Austin, Bureau of Economic Geology Report, in press.

Hubbard, R. J., Pape, J., and Roberts, D. G., 1985, Depositional sequence mapping to illustrate the evolution of a passive continental margin: In Seismic Stratigraphy II, Eds., O.R. Berg and D.G. Woolverton, AAPG Mem. 39, p.93-116.

Humphrey, J. D., 1988, Late Pleistocene mixing zone dolomitization, southeastern Barbados, West Indies: Sedimentology, v. 35, 327-348.

Humphrey, J. D. and Quinn, T. M., 1989, Coastal mixing zone dolomite, forward modelling, and massive dolomitization of platform-margin carbonates: Jour. Sedimentary Petrology, v. 59, 438-454.

Hunter, R. E., 1993, An eolian facies in the Ste. Genevieve Limestone of southern Indiana: in B. D. Keith and C.W. Zuppann, Mississippian Oolites and Modern Analogs, AAPG Studies in Geology #35, p. 163-174.

Hsu, K. J. and Schneider, J., 1973, Progress report on dolomitization - hydrology of Abu Dhabi sabkhas, Arabian Gulf, in B.H. Purser, eds., The Persian Gulf; Holocene Carbonate Sedimentation and Diagenesis in a Shallow Epicontinental Sea: New York, Springer-Verlag, p. 409-422.

Jacobs, D. K. and Sahagian, D. L., 1993, Climate-induced fluctuations in sea-level during non-glacial times: Nature, v. 361, p. 710-712.

Jankowsky, W. J. and Schlapak, G., 1983, Guyana Offshore, In Seismic Expression and Structural Styles, AAPG Studies in Geology Series # 15, Ed. A.W. Bally, p. 2.2.3-47 to 50.

James, N. P., 1983, Reefs, In Scholle, P.A. et al., Carbonate Depositional Environments. AAPG Mem. 33, p. 346-440.

James, N. P. and Choquette, P. W., eds., 1988, Paleokarst: Springer-Verlag, New York, p. 416.

James, N. P., 1994, Paleozoic cryocarbonates: Charlatans in the mist (abs.): Cool and Cold Water Carbonate Conference, Geelong, Victoria, Sedimentology Studies Group, Geol. Soc. Australia, p. 43.

Jordan, T. E. and Flemings, P.B., 1991, Large-scale stratigraphic architecture, eustatic variation and unsteady tectonism: A theoretical evaluation: Journal of Geophysical Research, v. 96, p. 6681-6699.

Kelleher, G. T. and Smosna, R., 1993, Oolitic tidal -bar reservoirs in the Mississippian Greenbriar Group of West Virginia: in B. D. Keith and C.W. Zuppann, Mississippian Oolites and Modern Analogs, AAPG Studies in Geology #35, p. 163-174.

Kendall, G. St. C. and Schlager, W., 1981, Carbonates and relative changes in sea level: Marine Geology ,v. 44, p. 181-212

Kerans, C., 1988, Karst-controlled reservoir heterogeneity in Ellenburger Group carbonates of West Texas: A.A.P.G. Bull., v. 72, p. 1160-1173.

Kerans, C. and Nance, H. S., 1991, High frequency cyclicity and regional depositional patterns of the Grayburg Formation, Guadalupe Mountains, New Mexico: In S. Meader-Roberts, M.P. Cadelaria and G.E. Moore (Eds), Sequence Stratigraphy, Facies and Reservoir Geometries of the San Andres, Grayburg, and Queen Formations, Guadalupe Mountains, New Mexico and Texas: Permian Basin Section SEPM Publication No. 91-32.

Kerans, C., Fitchen, W. M., Gardner, M. H., Sonnenfeld, M. D., Tinker, S. W., and Wardlaw, B. R., 1992, Styles of sequence development within uppermost Leonardian through Guadalupial strata of the Guadalupe Mountains, Texas and New Mexico: Permian Basin Exploration and Production Strategies p. 1-6.

Knight, I., James, N. P., and Lane, T. E., 1991, The Ordovician St. George unconformity, northern Appalachians: The relationship of plate convergence at the St. Lawrence Promontory to the Sauk/Tippecanoe sequence boundary: Geological Society of America Bull., v. 103, p. 1200- 1225.

Koerschner, W. F. and Read, J. F., 1989, Field and modelling studies of Cambrian carbonate cycles, Virginia Appalachians: Jour. Sed. Petrology, v. 59, p. 654-687.

Kominz, M. A., Beavan, J., Bond, C. G., and McManus, J., 1991, Are cyclic sediments periodic? Gamma analysis and spectral analysis of Newark Supergroup lacustrine strata: in E.K. Franseen, W.L. Watney, C.G.St.C. Kendall, and W.Ross, eds., Sedimentary Modelling, Kansas Geol. Surv. Bull., v. 233, p. 335-344.

Kozar, M. G., Weber, L. J., and Walker, K. R., 1990, Field and modelling studies of Cambrian carbonate cycles, Virginia Appalachians - Discussion: Jour. Sedimentary Petrology, v. 60, p. 790-794.

Kupecz, J. A., 1989, Petrographic and geochemical characterization of the Lower Ordovician Ellenburger Group, west Texas: Ph.D. Dissertation, University of Texas, Austin, 157 p.

Lees, A., 1975, Possible influence of salinity and temperature on modern shelf carbonate sedimentation Marine Geology, v. 19, p. 159-198.

Logan, B. W., 1974, Inventory of diagenesis in Holocene-Recent carbonate sediments, Shark Bay, Western Australia, in: B.W. Logan et al., Evolution and diagenesis of Quaternary carbonate sediments, Shark Bay, Western Australia: A.A.P.G. Memoir 22, p. 195-250.

Longacre, S. A., 1980, Dolomite reservoirs from Permian biomicrites: in R.B. Halley and R.G. Loucks, eds., Carbonate Reservoir Rocks, SEPM Core Workshop No. 1, p. 105-117.

Longman, M. W., Fertal, T. G., and Glennie, J. S., 1983, Origin and geometry of Red River Dolomite reservoirs, western Williston Basin: A.A.P.G. Bull. 67, p. 744-771.

Major, R. P., Bebout, D. G., and Lucia, F. J., 1988, Depositional facies and porosity distribution, Permian (Guadalupian) San Andres and Grayburg Formations, P.J.W.D.M. field complex, Central Basin Platform, West Texas: in A.J. Lomando and P.M. Harris, eds., Giant Oil and Gas Fields, A Core Workshop, SEPM Core Workshop 12, v. 2, p. 615-648.

Markello, J. R. and Read, J. F., 1982, Upper Cambrian intrashelf basin, Nolichucky Formation, southwest Virginia Appalachians: AAPG Bull., v. 66, p. 860-878.

Marquis, S. A. Jr., and. Laury, R. L, 1989, Glacio-eustasy, depositional environments, and reservoir character of Goen Limestone cyclothem (Desmoinesian), Concho Platform, Central Texas: A.A.P.G. Bull. 73, p. 166-181.

Martindale, W. and Boreen, T., 1995, Mississippian cool-water carbonate hydrocarbon reservoirs in the Southern Foothills of the Canadian Rocky Mountains (abs.): Cool and Cold Water Carbonate Conference, Geelong, Victoria, Sedimentology Studies Group, Geol. Soc. Australia, p. 46-47.

Matthews, M. D. and Perlmutter, M. A., 1994, Global cyclostratigraphy: an application to the Eocene Green River Basin. In Orbital Forcing and Cyclic Sequences, Eds., P.L. de Boer and D.G. Smith, Spec. Publ. No. 19, Int. Assoc. Sedimentologists, p. 459-482.

Matthews, R. K. and Frohlich, C., 1987, Forward modelling of bank-margin carbonate diagenesis: Geology, v. 15, p. 673-676.

Mazzullo, A. J. and Friedman, G. M., 1975, Conceptual model of tidal influenced deposition on margins of epeiric seas Lower Ordovician (Canadian) of eastern New York and southwestern Vermont: AAPG Bull., v. 59, p. 2123-2141.

McGowran, B., Li, Q., and Moss, G., 1991, The neritic carbonate record in southern Australia (abs.): in Cool and Cold Water Carbonate Conference, Sedimentology Studies Group, Geological Society of Australia, Geelong, Victoria, Australia, p. 49-50.

McKenzie, J. A., K. J. Hsu and J. F. Schneider, 1980, Movement of subsurface waters under the sabkha, Abu Dhabi, U.A.E., and its relation to evaporative dolomite genesis: SEPM Spec. Publ. No. 28, p. 11-30.

Meyer, Franz O., 1989, Siliciclastic influence on Mesozoic platform development: Baltimore Canyon Trough, Western Atlantic: Controls on Carbonate Platform and Basin Development, SEPM Spec. Publ. No. 44, p. 213-232.

Mitchell, J. C., Lehman, P. J., Cantrell, D. J., Al-Jallal, I. A., and Al-Thagafy, M.A.R., 1988, Lithofacies, diagenesis and depositional sequence: Arab-D Member, Ghawar Field, Saudi Arabia, in SEPM Core Workshop No. 12, p. 459-514.

Mitchum, R. M., Jr. and Van Wagoner, J. C., 1991, High frequency sequences and their stacking patterns: sequence-stratigraphic evidence of high-frequency eustatic cycles: Sedimentary Geology, v. 70, p. 131-160.

Montanez, I.P. and Osleger, D.A., 1993, Parasequence stacking patterns, third-order accommodation events, and sequence stratigraphy of Middle to Upper Cambrian platform carbonates, Bonanza King Formation, southern Great Basin: in R.G. Loucks and J. Frederick Sarg, eds., Carbonate Sequence Stratigraphy, AAPG, p. 305-326.

Montanez, I. P. and Read, J. F., 1992a, Fluid-rock interaction history during stabilization of early dolomites, Upper Knox Group (Lower Ordovician), U.S. Appalachians: Jour. Sedimentary Petrology, v. 62, p. 753-778.

Montanez, I. P. and Read, J. F., 1992b, Eustatic control on dolomitization of cyclic peritidal carbonates: Evidence from the Early Ordovician Knox Group, Appalachians: Geological Society of America Bull., v. 104, p. 872-886.

Mundil, R., Brack, P., Meier, M., and Oberli, F., in press, Calibration of Triassic "Milankovitch cycles" by high resolution U-Pb age determination: IAS.

Mussman, W. J., Montanez, I. P., and Read, J. F., 1988, Ordovician Knox paleokarst unconformity, Appalachians: in N.P. James and P.W. Choquette, eds., Paleokarst, Springer-Verlag, New York, p. 211-228.

Niemann, J. C. and Read, J. F., 1988, Regional cementation from unconformity-recharged aquifer and burial fluids, Mississippian Newman Limestone, Kentucky: Jour. Sediment. Petrology, v. 58, p. 688-705.

Nelson, W. A. and Read, J. F., 1989, Updip to downdip cementation and dolomitization patterns in a Mississippian aquifer, Appalachians: Jour. Sedimentary Petrology, v. 60, p. 379-396.

Osleger, D. and Read, J. F., 1991, Relation of eustasy to stacking patterns of meter-scale carbonate cycles, Late Cambrian, U.S.A.: Journal of Sedimentary Petrology, v. 61, No. 7, p.1225-1252.

Perkins, R. D., 1977, Depositional framework of Pleistocene rocks in South Florida: in P. Enos and R. Perkins, Quaternary Sedimentation in South Florida, The Geological Society of America Memoir 147, p. 131-198.

Pomar, L., 1991, Reef geometries, erosion surfaces and high-frequency sea-level changes, Upper Miocene Reef Complex, Mallorca, Spain, Sedimentology, v. 38, p. 243-269.

Pomar, L. N., 1993, High-resolution Sequence Stratigraphy in Prograding Miocene Carbonates: Application to Seismic Interpretation, in R.G. Loucks and J. Frederick Sarg, eds., Carbonate Sequence Stratigraphy, AAPG, p. 389-434.

Posamentier, H. W. and Allen, G. P., 1993, Siliciclastic sequence stratigraphic patterns in foreland ramp-type basins: Geology, v. 21, p. 455-458.

Posamenteir, H. W., Summerhayes, C. P., Hag, B. U., and Allen, G. P. (Editors), 1993. Sequence stratigraphy and facies, Int. Assoc. Sedimentol., Spec. Publ., 18 (in press).

Pray, L. C. and Wray, J. L., 1963, Porous algal facies (Pennsylvanian) Honaker Trail, San Juan Canyon, in R.O. Bass, ed., Four Corners Geological Society, 4th Field Conf. Gdbk., p. 204-234.

Purdy, E. G., 1974a, Karst-determined facies patterns in British Honduras: Holocene carbonate sedimentation model, Bull. Amer. Assoc. Petrol. Geol., 58:825-855.

Purdy, E. G., 1974b, Reef configurations: cause and effect, In: Laporte, L.E. (editor), Special Publication of the Society of Economic Paleontologists and Mineralogists, 18:9-76.

Read, J. F., 1973, Paleo-environments and paleogeography, Pillara Formation (Devonian), Western Australia: Bull. Canadian Petroleum Geology, v. 21, p. 344-394.

Read, J. F., 1985, Carbonate platform models: AAPG Bull., v. 69, p. 1-21.

Read, J. F. and Goldhammer, R. K., 1988, Use of Fischer plots to define 3rd order sea-level curves in peritidal cyclic carbonates, Early Ordovician, Appalachians: Geology, v. 6, p. 895-899.

Read, J. F. and Horbury, A. D., 1993, Eustatic and tectonic controls on porosity evolution beneath sequence-bounding unconformities and parasequence bounding disconformities on carbonate platforms: in A.D. Horbury and A.G. Robinson, Eds., Diagenesis and Basin Development, AAPG Studies in Geology #36, p. 155-197.

Read, J. F., Osleger, D. A., and Elrick, M. E., 1991, Two-dimensional modelling of carbonate ramp sequences and component cycles: in E.K. Franseen et al., eds., Sedimentary Modelling: Computer simulations and methods for improved parameter definition: Kansas Geological Survey Bull. 233, p. 473-488.

Reid, A. M. and Tomlinson-Reid, S. A., 1991, The Cogdell field study, Kent and Scurry Counties, Texas: A post-mortem: in M.P. Candelaria, ed., Permian Basin Plays - Tomorrow's Technology Today, West Texas Geological Society Publ. No. 91-89, p. 39-66.

Roylance, M. H., 1990, Depositional and diagenetic history of a Pennsylvanian algal-mound complex: Bug and Papoose Canyon fields, Paradox Basin, Utah and Colorado: A.A.P.G. Bull. 74, p. 1087-1099.

Ruddiman, W. F., Raymo, M., and McIntyre, A., 1986, Matuyama 41,000-year cycles: North Atlantic Ocean and northern hemisphere ice sheets, Earth and Planetary Science Letters, v. 80, p. 117-129

Ryder, R., Harris, A. G., and Repetski, J. E., 1992, Stratigraphic framework of Cambrian and Ordovician rocks in the Central Appalachian Basin from Medina County, Ohio, through Southwestern and South-Central Pennsylvania to Hampshire County, West Virginia: U.S. Geol. Survey Bull. 1839, Ch. K, 32 p.

Sadler, P. M., Osleger, D. A., and Montanez, I. P., 1993, On labeling, interpretation, and the length of Fischer plots, Jour. Sed. Petrology, v. 63, p. 360-368.

Saller, A. H., 1984, Petrologic and geochemical constraints on the origin of subsurface dolomite, Enewetak Atoll: an example of dolomitization by normal seawater: Geology, v. 12, p. 217-220.

Sarg, J. F. and Lehmann, P. J., 1986, Lower-Middle Guadalupian facies and stratigraphy, San Andres/Grayburg Formations, Permian Basin, Guadalupe Mountains, New Mexico: Permian Basin Section SEPM Publ. No. 86-25, p. 1 to 35.

Sarg, J. F., 1988, Carbonate Sequence Stratigraphy: in C.K. Wilgus and others, eds., Sea-Level Changes: An Integrated Approach: SEPM Spec. Publ. 42, p. 156-181.

Schatzinger, R. A., 1983, Phylloid algal and sponge-bryozoan mound-to-basin transition: A Late Paleozoic facies tract from the Kelly-Snyder field, West Texas: in P.M. Harris, ed., Carbonate Buildups - A Core Workshop, SEPM Core Workshop No. 4, p. 244-303.

Schlager, W., 1981, The paradox of drowned reefs and carbonate platforms: Geological Society of American Bulletin, v.92, p. 197-211.

Schlager, W., 1989, Drowning unconformities on carbonate platforms: Controls on Carbonate Platform and Basin Development, SEPM Special Publication No. 44, p. 15-25.

Schlager, W., 1992, Sedimentology and sequence stratigraphy of reefs and carbonate platforms: AAPG Continuing Education Course Note Series #34, 71 p.

Schwarzacher, W. and Haas, J., 1986, Comparative statistical analysis of some Hungarian and Austrian Upper Triassic peritidal carbonate sequences: Acta Geologica Hungarica, v. 29, p. 175-196.

Sloss, L. L., 1963, Sequences in the cratonic interior of North America: Geol. Soc. America Bull., v. 74, p. 93-113.

Smart, P. L., Palmer, R. J., Whitaker, F., and Wright, V. P., 1988, Neptunian dikes and fissure fills: an overview and account of some modern examples: in N.P. James and P.W. Choquette, eds., Paleokarst, Springer-Verlag, New York, p. 149-   163.

Smith, T. and Dorobek, S. L., 1993, Alteration of early-formed dolomite during shallow to deep burial:

Mississippian Mission Canyon Formation, central to southwestern Montana: Geol. Soc. America Bull., v. 105, p. 1389-1399.

Somerville, I. D., 1979a, Minor sedimentary cyclicity in late Asbian (upper D1) limestones in the Llangollen district of North Wales: Proc. Yorkshire Geol. Soc., v. 42, p. 317-341.

Somerville, I. D., 1979b, A cyclicity in early Brigantian (D2) limestones east of the Clwydian Range, North Wales, and its use in correlation: Geological Journal, v. 14, p. 69-86.

Sonnenfeld, M. D., 1991, High cyclicity within shelf-margin and slope strata of the Upper San Andres sequence, Last Chance Canyon: In Sequence Stratigraphy, Facies, and Reservoir Geometries of the San Andres, Grayburg, and Queen Formations, Guadalupe Mountains, New Mexico and Texas. Permian Basin Section-SEPM Ann. Field Trip, Eds., S. Meader-Roberts, M.P. Candelaria, and G.E. Moore, Permian Basin SEPM Publ. 91-32, p. 11-52.

Strasser, A., 1988, Shallowing-up sequences in Purbeckian peritidal carbonates (lowermost Cretaceous, Swiss and French Jura Mountains): Sedimentology, v. 35, p; 369-383.

Todd, R. G. and Mitchum, R. M. Jr., 1977, Identification of Upper Triassic, Jurassic and Lower Cretaceous Seismic Sequences in Gulf of Mexico and Offshore West Africa: In Seismic Stratigraphy - Applications to Hydrocarbon Exploration, Ed. C.E. Payton, AAPG Mem26, p. 145-163.

Tucker, M. E., 1985, Shallow-marine carbonate and facies models, in P.J. Benchley and B.P.J. Williams, Sedimentology: recent developments and applied aspects, Geol. Soc. London Spec. Publication 18, p. 139-161.

Vail, P. R., 1987, Seismic stratigraphic interpretation using sequence stratigraphy. Part 1: Seismic stratigraphy interpretation procedure, in: Bally,A.W. (ed), AAPG Atlas of Seismic Stratigraphy, v. 1, AAPG Studies in Geology, no. 27, p. 1-10.

Vail, P. R., Audemard, F., Bowman, S. A., Eisner, P. N. and Perez-Cruz, C., 1991, The stratigraphic signatures of tectonics, eustasy and sedimentology - an overview: in G. Einsele, W. Ricken and A. Seilacher, eds., Cycles and Events in Stratigraphy, Springer-Verlag, p. 617-659.

Vail, P. R., Mitchum, R. M., and Thompson, S., III, 1977, Seismic stratigraphy and global changes of sea level, Part 4: Global cycles of relative changes of sea level, in: C.E. Payton, ed., Seismic Stratigraphy - Applications to Hydrocarbon Exploration, AAPG Memoir 26, p. 83-97.

Van Hinte, J. E., 1978, Geohistory analysis - application of micropaleontology in exploration geology: AAPG Bull., v. 62, p. 201-222.

Van Wagoner, J. C., Posamentier, H. W., Mitchum, R. M., Vail, P. R., Sarg, J. F., Loutit, T. S., and Hardenbol, J., 1988, An overview of the fundamentals of sequence stratigraphy and key definitions; in C.K. Wilgus et al., eds., Sea-Level Changes - An Integrated Approach, SEPM Spec. Publ. 42, p. 39-46.

Walkden, G. M., 1987, Sedimentary and diagenetic styles in late Dinantian carbonates of Britain, in: J. Miller, A.E. Adams and V.P. Wright, eds., European Dinantian Environments. Wiley, Chichester, p.131-155.

Walkden, G. M. and Walkden, G. D., 1990, Cyclic sedimentation in carbonate and mixed carbonate-clastic environments: four simulation programs for a desktop computer: in M.E. Tucker et al., eds., Carbonate Platforms, Facies Sequences and Evolution, Spec. Publ. No. 9, International Assoc. of Sedimentologists, p. 55-78.

Walls, R. A. and Burrowes, G., 1985, The role of cementation in the diagenetic history of Devonian reefs, western Canada, in N. Schneidermann and P.M. Harris, eds., Carbonate Cements: SEPM Spec. Publ. 36 p. 185-220.

Ward, W. C. and Halley, R.B., 1985, Dolomitization in a mixing zone of near-seawater composition, Late Pleistocene, northeastern Yucatan Peninsula: Jour. Sedimentary Petrology, v. 55, p. 407-420.

Watney, W. L., French, J., and Franseen, E. K., 1989, Sequence stratigraphic interpretations and modelling of cyclothems in the Upper Pennsylvanian (Missourian) Lansing and Kansas City groups in eastern Kansas, Kansas Geological Society 41st Annual Field Trip Guidebook, Lawrence, Kansas, 211 p.

Watney, W. L., Wong, J-C, and French, J. A. Jr., Computer simulation of Upper Pennsylvanian (Missourian) carbonate-dominated cycles in western Kansas: in E.K. Franseen, W.L. Watney, C.G.St.C. Kendall, and W.Ross, eds., Sedimentary Modelling, Kansas Geol. Surv. Bull., v. 233, p. 415-430.

Watts, A. B., 1981, The U.S. Atlantic continental margin: subsidence history, crustal structure and thermal evolution: in A.W. Bally et al., eds., Geology of Passive Continental Margins, AAPG Education Course Note Series #19, p. 2-1 - 2-75.

Wendte, J. C., 1992, Cyclicity of Devonian strata in the Western Canada Sedimentary Basin: In Wendte, J., Stoakes, F.A. and Campbell, C.V., Devonian-Early Mississippian carbonates of the Western Canadian Sedimentary Basin: A Sequence Stratigraphic Framework, SEPM Short Course No. 28, Calgary, p. 25-39.

Wendte, J. C., 1992, Evolution of the Judy Creek complex, a Late Middle Devonian isolate platform-reef complex in west-central Alberta, In Wendte, J.C. et al., Devonian-Early Mississippian carbonates of the Western Canada Sedimentary Basin: A Sequence Stratigraphic Framework. SEPM Short Course No. 28, Calgary, p. 89-125.

Wendte, J. C. and Stoakes, F. A., 1982, Evolution and porosity development of the Judy Creek reef complex, Upper Devonian, Central Alberta: in W.G. Cutler, Canadas Giant Hydrocarbon Reservoirs, Canadian Society of Petroleum Geologists, Calgary, Alberta, p. 63-81.

Whittaker, F. F. and Smart, P. L., 1990, Active circulation of saline ground waters in carbonate platforms: Evidence from the Great Bahama Bank: Geology, v. 18, p. 200-203.

Wilson, J. L., 1975, Carbonate facies in geologic history: Springer-Verlag, New York, 471 p.

Wright, V. P., 1992, Speculations on the controls on cyclic peritidal carbonates: ice-house versus greenhouse eustatic controls: Sedimentary Geology, v. 76, p. 1-5.

Wright, V. P., Estaban, M., and Smart, P., 1991, Palaeokarst and palaeokarstic reservoirs: University of Reading.

Wright, V. P. and Faulkner, T. J., 1990, Sediment dynamics of early Carboniferous ramps: Geol. Jour., v. 25, p. 139-144.

PART 2

USE OF ONE- AND TWO-DIMENSIONAL CYCLE ANALYSIS IN ESTABLISHING HIGH-FREQUENCY SEQUENCE FRAMEWORKS

by Charles Kerans

USE OF ONE- AND TWO-DIMENSIONAL CYCLE ANALYSIS IN ESTABLISHING HIGH-FREQUENCY SEQUENCE FRAMEWORKS

CHARLES KERANS
Bureau Of Economic Geology
The University of Texas at Austin
Austin, Texas 78713

INTRODUCTION

The central goal of this contribution is to development new methods for describing and modeling carbonate reservoirs. This pair of exercises focuses on establishing a high-resolution sequence framework for shallow-water platform-top carbonate reservoir strata and, for these specific examples, mixed-carbonate-clastic successions. The scale of particular interest is that of several thousand feet laterally and several hundred feet vertically, a scale useful in reservoir description and modeling.

Specific topics for this discussion will be the development of a sequence stratigraphic framework that integrates observations from core, wireline logs, and outcrop analogs. This approach is similar to that applied by Van Wagoner et al (1988, 1990) for siliciclastic reservoirs. It downplays the emphasis on stratal geometry, the dominant player in seismically oriented sequence stratigraphy, putting it on par with other sequence stratigraphic tools, including cycle hierarchy, cycle stacking (both thickness and symmetry), facies proportions, and facies tract offset.

The data for the two exercises used in this segment of the course are from the Queen Formation outcrops on the Shattuck Escarpment in the Central Guadalupe Mountains, and Grayburg Formation core from the North Cowden Grayburg reservoir of the Central Basin Platform, West Texas (Figs. 1, 2, 3). Both these data sets provide an excellent opportunity for geological exercises focused on development of high-frequency sequence frameworks useful in reservoir characterization. Favorable attributes of these data sets are (1) a relatively thin and manageable genetic stratigraphic interval (250-350 ft) that contain hierarchical levels of cyclicity, (2) abrupt dip-related facies changes within discrete stratigraphic intervals making high resolution correlation challenging, (3) a relatively simple suite of 9 macroscopically identifiable depositional facies that are sufficient to describe the sequences contained within these facies down to a sufficiently fine level of detail without the aid of petrographic examination.

These attributes will allow us to spend relatively little time on logistics or routine facies examination, and a maximum amount of time on recognition of patterns and correlations of high-frequency cyclicity and its use as a tool for development of cycle-scale sequence stratigraphic reservoir frameworks.

TERMINOLOGY OF HIGH FREQUENCY SEQUENCE STRATIGRAPHY

Extensive analysis of the Permian outcrops of the Guadalupe Mountain area of West Texas and New Mexico, and the adjoining subsurface reservoirs has led to the conclusion that the fundamental level for establishing a high-resolution sequence framework is that of genetic cycles and/or cycle sets (depending on the accommodation setting)(Kerans et al., 1994). Recognizing and describing the interval of interest at this scale of the high-frequency cycle makes it possible to generate superior layer models for reservoir description, volumetric calculations, and fluid flow modeling that do not arbitrarily average out critical heterogeneities. Further, facies analysis within this high-frequency chronostratigraphic scale provides a means of developing depositional models that are scale-sensitive for use in forward modeling and stochastic modeling of reservoir strata.

The two exercises will stress the observation, description, and interpretation of cycles in both 1 and 2 dimensions (measured sections and cross sections) in mixed siliciclastic-carbonate settings. A background of basic concepts and methods of cycle stratigraphy and its integration into sequence stratigraphy will be needed for background.

The terminology of high-frequency sequence stratigraphy used in here is based on Mitchum and Van Wagoner's (1991) modification of the Exxon-type sequence stratigraphic terminology (Vail, 1987; Van Wagoner et al., 1988) (Fig. 3). This terminology emphasizes the composite eustatic signal that controls the development of the San Andres, and most other carbonate shelf-to-basin stratigraphic successions. Composite sequences are made up of high-frequency sequences (HFS), which in turn consist of high-frequency cycles (cycles).

USE OF ONE- AND TWO-DIMENSIONAL CYCLE ANALYSIS

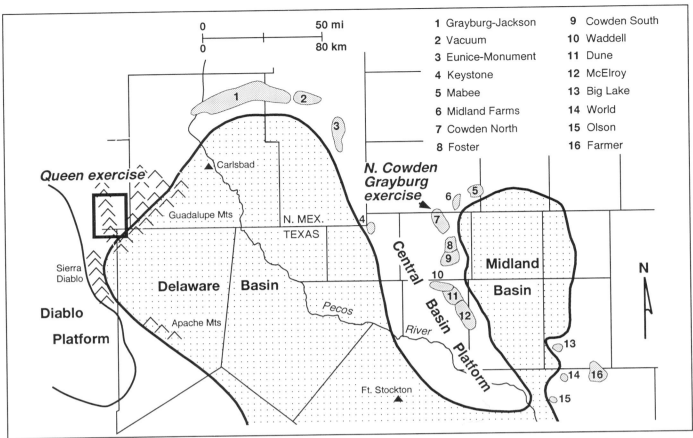

Figure 1. Location map showing general paleogeography of Permian Basin during the Guadalupian and position of oil reservoirs in the Grayburg Formation that have produced greater than 10 MMbbl. The two exercise data sets from the Queen Formation outcrop in the Guadalupe Mountains, and the Grayburg Formation at North Cowden Field are given. A detail of the Guadalupe Mountain area is given in figure 7.

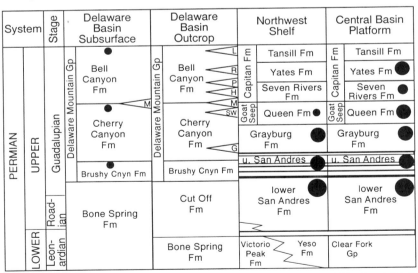

Figure 2. Stratigraphic terminology of the Guadalupian section of the Permian Basin, including the Grayburg and Queen formations and their basinal equivalents.

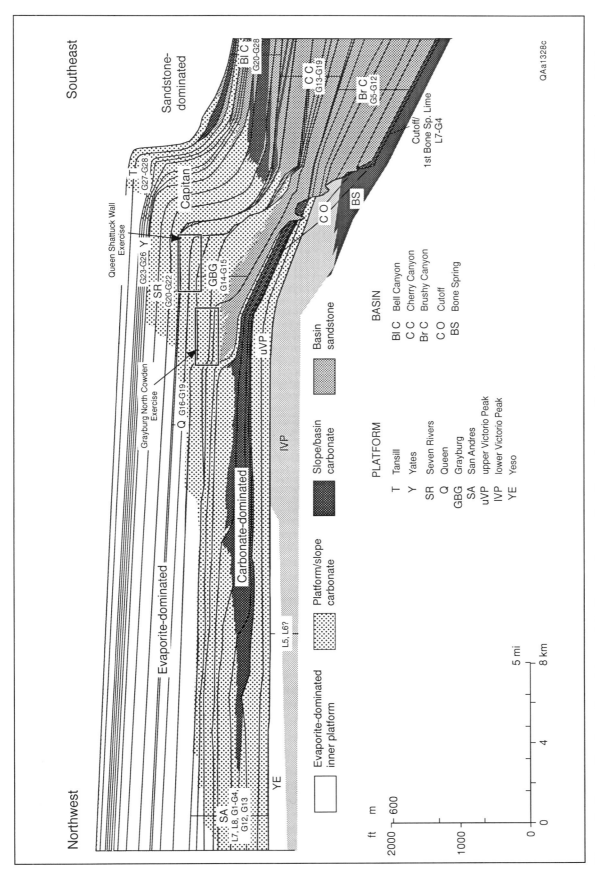

Figure 3. Generalized stratigraphic model of upper Leonardian through upper Guadalupian sequences in the Guadalupe Mountains and Delaware Basin, Texas and New Mexico. Gross relationships between evaporite, shoal-water carbonate, deep-water carbonate, and basinal sandstone are illustrated. Sequence abbreviations include a letter for age (L=Leonardian, G=Guadalupian) and a number for relative position within a series. Broad carbonate ramps with decoupled basinal siliciclastics typify uppermost Leonardian-lower Guadalupian sequences (L7-G13). Significant contraction of carbonate facies tracts, increased preservation of siliciclastics on the platform, and coupled siliciclastic-carbonate facies tracts at intermediate and high-order cycle scales characterize mid-upper Guadalupian sequences (G14-G28). Stratigraphic position of data sets for course exercises is shown.

Both high-frequency and composite sequences conform to the general definition of depositional sequences, being relatively conformable sets of strata bounded by subaerial unconformities or their correlative conformities. Composite sequences are most similar in scale to the seismically resolvable depositional sequences originally discussed by Vail and co-workers, but differ in that they contain high-frequency sequences and their associated unconformities. High-frequency sequences stack into lowstand, transgressive, and highstand sequence sets that definine the systems tracts of composite sequences, whereas cycles occur in retrogradational, aggradational, and progradational sets, defining the systems tracts of high-frequency sequences.

The term "high-frequency cycle" is substituted for parasequence in this terminology because it stresses the parallel cyclic hierarchy of cyclicity, whereas parasequences are thought to be asymmetrical progradational depositional events recording only high-frequency base-level fall time. In carbonate systems, times of base-level rise may result in great or greater sedimentation, and thus a symmetrical record of base-level rise and fall is possible.

This nested or composite stratigraphic cyclicity in the San Andres and equivalent strata of the Permian Basin was first recognized by Miessner (1970) and was subsequently emphasized by Kerans et al. (1992) and Gardner (1992) in their analyses of the Guadalupian sequence framework. A similar emphasis on recognizing the composite-cyclic framework of stratigraphic successions has been presented by Koershner and Read (1989) and Goldhammer et al. (1990) among others. An attempt has been made here to downplay the labeling of the hierarchy of stratigraphic cycles in terms of orders (for example 3rd, 4th, 5th) as the rigorous chronologic analysis required to properly assign these time-specific terms has not been carried out. However, an estimate would be; cycles=5th, high-frequency sequences=4th, and composite sequences=3rd order.

Tools of Sequence Stratigraphy

The tools listed below are summarized from work by Gardner (Gardner et al., 1992) and the genetic stratigraphy group at Colorado School of Mines supervised by T. A. Cross as well as by other workers. This list of observations provides a systematic checklist of observations that can be applied in the analysis of any stratigraphic succession where it is considered desirable to understand the longer term stratigraphic influences on the sedimentologic, and hence, reservoir character.

Cycle stacking.—

Stacking pattern analysis examines a stratigraphic section for systematic upward changes in cycle thickness. The Fischer plot (Read and Goldhammer, 1988) is a tool for graphically analyzing a section for cycle thickening and thinning patterns, as normalized to a constant subsidence. In an accommodation-limited setting such as a flat-topped carbonate platform, or in a system where storm or fair-weather wave base serves as an accommodation limited, an upward thinning of cycles generally signals fill-up of accommodation, with the position of thinnest cycles marking the approximate position of a sequence boundary. Upward thickening trends may be characteristic of transgressive systems where accommodation space is increasing. This technique is a simple and quick way to examine a large amount of data rapidly and can be useful as a correlation tool. However, it represents only one of several tools, and its results generally do not lead to unique solutions. For example, upward thinning of cycles can reflect decreased sediment supply as well as decreased accommodation. In carbonate successions, during a rapid eustatic rise, cycle thinning may occur when water depths exceed 100 ft where sediment production rates decrease.

Facies Proportions.—

Stacking pattern analysis by itself may lead to a non-unique solution, but if performed in combination with analysis of facies proportions, a clear picture of accommodation trends and changes in platform facies evolution will result. In the case where upward cycle thinning is reflecting decreasing accommodation associated with relative sea-level fall and sequence boundary formation, one should also observe a increase in the ratio of peritidal facies to subtidal facies in each successive cycle in concert with cycle thinning. Increasing dolomitization is also commonly associated with this trend. In contrast, upward thinning of cycles in the later portion of the transgressive systems tract (the result of sediment starvation) might be confused on a well-log cross section can be easily differentiated by core or cuttings data, or log facies patterns that show an increase in deeper, lower energy subtidal facies. Thus the combination of information from facies proportion and cycle thickness will allow solution of this problem.

Cycle Symmetry.—

A third method for describing vertical depositional trends is to divide for every cycle the portion that records transgression or base-level rise from the portion that records regressive or base-level fall. In our example from the Queen Formation, on the platform top the transgressive hemicycle may be thin and difficult to resolve in most cycles. However it is the exception where in a cycle preserves a thick transgressive hemicycle that may be the most important for correlation, as well as for forming continuous reservoir baffles (or even possibly thief zones depending on facies type and diagenesis).

Stratal Preservation—

An observation that is commonly overlooked is the relative importance of diastems or minor erosional surfaces in stratigraphic successions. These surfaces of erosion or non-deposition generate considerable discomfort for most geologists. The common approach is to assume if they are not described they do not exist. Rather, the presence and distribution of erosion and-or truncation surfaces and abrupt facies transitions should be noted where observed, regardless of the fact that the time represented by the surface is not known. The distribution of erosive surfaces, when plotted alongside criteria described above, should yield an additional independent observational indicator of trends in creation or loss of accommodation and resultant facies change.

Facies tract offset.—

Once a general depositional model has been established that places facies observed in vertical successions into a lateral, Waltherian, perspective, it is possible to focus on perturbations from an idealized succession of facies. For example, assume a prograding carbonate depositional ramp model with ramp-crest grainstone, outer ramp skeletal wackestone, and distal ramp mudstone distributed coevally from shelf to basin. A measured section of this succession would be expected to contain mudstone passing upwards into wackestone and grainstone. If however, grainstone is found to rest directly on distal outer ramp mudstone, then this would signal a key facies tract offset, in this particular case being a downward shift in coastal onlap. Such "out of sequence" shifts in vertical facies successions point to critical events and surfaces that most likely will be associated with abrupt changes in reservoir quality and/or continuity.

Stratal Geometry.—

Observations on the scale of stratal geometries typically require large 2 or 3-dimensional views such as are available on seismic data or in large outcrops. However, these observations are critical as they form the basis for complex correlation schemes that utilized inclined time-lines that might be unregonized in the course of standard subsurface well log correlation. Systematic changes in stratal geometry form the core of the working models of sequence stratigraphy, having been the basis of the Vail et al. (1977) initial stratigraphic model. Critical geometric relationships are now widely standardized in terms of onlap, downlap, and erosional truncation stratal terminations, presence or absence of clinoforms and their oblique versus sigmoid character, and a range of other seismic facies parameters. Carbonate stratigraphic sequences commonly have stratal geometries that are closely comparable to their siliciclastic counterparts, but strong divergences between these systems can exist as a result of the tendency of carbonate sediments to be locally sourced as well as the ability of some carbonate systems to keep pace with rapid sea-level rise. The most striking difference between carbonate and siliciclastic sequences involves their response to lowstand conditions. Where siliciclastic sequences commonly develop significant lowstand systems tract deposits during a sea-level fall, lowstand conditions shut down the carbonate sediment factory and thus may impart only a minimal lowstand record. Hanford and Loucks (1994) provide a recent review of some of the key stratal geometries in carbonate sequences.

Correlation of upward shallowing cycles and sets of cycles—

Analysis of cycle stacking and recognition of cycle sets in detailed measured sections, core, and logs allows the ranking of correlation surfaces. Those surfaces that are both tops of cycles, and of sets of cycles typically provide a higher reliability for correlation. Regonizing the turn-around from regressive to transgressive cycle stacking can be very useful in tracking surfaces across different lithofacies. Once cycles and cycle sets are recognized, key correlation horizons such as flooding surfaces at the base of cycle sets with the greatest facies tract offset can be targeted and traced.

In many examples with low accommodation such as the lower-middle Permian Clear Fork Formation of the Permian Basin, individual cycles can rarely be traced and cycle sets must be correlated (Ruppel, 1992).

Recognition of cycle boundaries of different hierarchical levels.—

In our Queen Formation exercise we will observe the complex nesting of several orders of cyclicity. It is critical to describe and compare all cycle patterns both within a section and between sections to assure that one is not mixing levels of cyclicity. Careful analysis of stacking patterns, facies composition, and magnitude of facies tract offset across cycle-bounding surfaces are keys to identifying differing levels of cyclicity.

Recognition of sequence boundaries vs. cycle boundaries —

Sequence boundary recognition may be relatively simple in the case of a major karst-modified, tectonically enhanced setting. However, in many if not most cases, it is a complex and highly interpretive procedure. Karstification, stratal truncation, and major downward shifts in coastal onlap (facies tract offset) are the most reliable indicators of sequence boundaries. Changes in stacking pattern from upward thinning cycles with increasing peritidal/subtidal facies ratios to upward thickening cycles with decreasing peritidal/subtidal facies ratios brackets the position of a sequence boundary.

In many examples sequence boundaries are distinct in one facies tract and poorly developed in another. In these cases it may be necessary to utilize several lines of observation. In most analyses of platform-top carbonate strata, the identification of the sequence boundary is of secondary importance relative to the knowledge of the detailed facies distribution.

Grouping of cycle types into facies tracts.—

Once cycle analysis has been carried out for individual measured sections, cores, or logs, and a hierarchical arrangement of cycles recognized and correlated, a reasonable chronostratigraphically significant framework exists that can be used as the basis for more detailed facies and depositional environmental analysis. Chronostratigraphically equivalent cycles from seperate measured sections can be compared in terms of facies composition, and facies tracts delineated. Ultimately a dynamically evolving depositional model can be derived from within this framework.

GENERAL STRATIGRAPHIC FRAMEWORK OF GUADALUPIAN SEQUENCES

San Andres Formation composite and high-frequency sequences —

Leonardian through Guadalupian stratigraphy of the Guadalupe Mountains can be divided into a series of 28 "high-frequency" sequences (Fig. 3). Our exercises from the Queen and Grayburg formations come from the Guadalupian 14 and 15 HFS (Grayburg Fm.) and Guadalupian 16 and 17 HFS (Queen Fm.).

The lowest sequences in this stack are Leonardian 7 and 8 and Guadalupian 1 through Guadalupian 13 (San Andres, Cutoff, Brushy Canyon, and Cherry Canyon Formations). These first 13 sequences make up two composite sequences (Leonardian 7 through Guadalupian 4 = lower San Andres composite sequence; Guadalupian 5 through Guadalupian 13 = upper San Andres composite sequence). The lower San Andres composite sequence is virtually devoid of siliciclastic facies, whereas the upper San Andres composite sequence is approximately 50% siliciclastics.

The distribution of siliciclastics in the upper San Andres composite sequence is decoupled at the high-frequency scale, becoming important mainly at the scale of the composite sequence. The basal 6 HFS compose an entirely basin-restricted siliciclastic lowstand sequence set, whereas the subsequent transgressive HFS set is dominantly of carbonate. Only the highstand HFS of this upper San Andres composite sequence, the Guadalupian 13 HFS, displays high-frequency siliciclastic/carbonate cyclicity (Sonnenfeld and Cross, 1994).

The lower San Andres composite sequence is associated with the great majority of hydrocarbon production in the Permian of the Permian Basin. These broad low-relief carbonate ramps dominated by generally dolomitized grain-supported rocks form moderate to high-quality reservoirs. Top and updip seals are formed by seaward overstepping evaporitic inner-ramp facies tracts.

Grayburg Formation (Artesia Group) high-frequency sequences.—

Guadalupian sequences 14 and 15 include the Grayburg Formation and its basinal equivalents in the Cherry Canyon Formation. These two sequences are effectively coupled with moderately thick siliciclastic-dominated lowstand deposits and transgressive and highstand tracts that average 50% siliciclastic sediments. Clinoforms in shelf-margin portions of these sequences have 10-15 degree slopes and youngest strata contain megabreccias that contain some reefal material. Little can be said concerning the character of the margin of the younger of the two Grayburg sequences as it has been truncated by a major collapse scar (Franseen et al., 1989).

Production from Grayburg reservoirs is predominantly from siliciclastics in the western portion of the producing trend whereas farther from the source of siliciclastics, reservoir facies are subtidal moderate to low-energy grain-supported carbonates. Grayburg HFS are the oldest Guadalupian platforms to develop distinct shelf margin and shelf crest facies tracts.

Queen Formation (Artesia Group) high-frequency sequences.—

The Queen Formation is made up of two HFS that can be recognized both in the outcrop area and in subsurface reservoirs throughout the Permian Basin. These sequences, Guadalupian 16 and 17, contain the first evidence of reef-rimmed platform development in the partially coeval Goat Seep Formation (Crawford, 1981) and are coupled sequences in the sense that both transgressive and highstand systems tracts display clastics and carbonates interbedded at the high-frequency (meter) scale. Further, highstand toe-of-slope facies of the South Wells member of the Cherry Canyon Formation also show high-order siliciclastic and carbonate alternations (King, 1948), indicating that siliciclastic bypass of the shelf to the basin was occurring throughout deposition of Queen strata on the shelf.

Production from Queen Formation reservoirs is almost entirely from siliciclastics in a middle shelf position.

Division into a lower and an upper productive sandstone by tight transgressive-systems-tract carbonates of the upper Queen sequence is common in many Queen reservoirs.

Seven Rivers, Yates, and Tansill high-frequency sequences.(Artesia Group)

An estimated 3 Seven Rivers sequences, 5 Yates sequences, and two Tansill sequences (Guadalupian 18-28) make up the remainder of the Artesia Group and equivalent basin facies of the Bell Canyon Formation. These sequences are comparable in depositional style to those initiated during upper Queen sedimentation with coupled high-frequency siliciclastic-carbonate reciprocal sedimentation patterns and well developed reef-rimmed margin, albeit deeper rimmed in earlier sequences. Siliciclastics are not distributed evenly throughout these sequences. Seven Rivers sequences show a decrease in clastic input relative to the preceding Queen sequences and subsequent Yates sequences. Meissner (1970) recognized this lower order cyclicity in siliciclastic deposition (San Andres- sand poor; Grayburg-Queen, sand rich, Seven Rivers sand poor, Yates Fm sand rich; Tansill sand poor) and inferred that this level of cyclicity could be traced northwards through the Permian Basin into the adjacent Permian strata of the Rocky Mountains.

Similarities in facies tract development between the Queen Formation and the overlying Seven Rivers through Tansill sequences are strong, but key differences exist. With successively younger sequences, the width of facies tracts decreases systematically, such that the width of the shelf-crest facies tract during Queen deposition is somewhat broader than that of the Tansill shelf-crest tract in Dark Canyon. Additional differences include more dramatic development of the teepee-pisolite facies in the Seven Rivers-Tansill sequences, better developed reef communities, thicker slope aprons of reef-derived talus, and far less fabric-destructive dolomitization.

LITHOSTRATIGRAPHIC BACKGROUND OF GRAYBURG AND QUEEN UNITS

The Grayburg and Queen Formations are sufficiently similar that a single set of depositional facies, facies tracts, and depositional models is provided below. This generalized lithostratigraphic characterization served only as a starting point. The four high-frequency sequences that compose these formations were deposited over a time range of approximately 2 million years, and the changing character of basin subsidence rates, sediment supply, climate, and eustacy clearly influenced the stratigraphic and sedimentologic attributes of these sequences.

Depositional Facies

The initial step in any stratigraphic analysis is to accurately describe vertical and lateral distributions of facies. During this initial phase, one needs to consider the grouping of these facies into genetic associations. Only through this approach will generalities, correlations and construction of a cycle-scale framework develop. The Grayburg and Queen mixed siliciclastic/carbonate shelf systems are sufficiently similar that a general set of facies can be distinguished for both

Massive quartzose siltstone-sandstone .—

A common facies in the Grayburg Formation, and the most volumetrically important facies in the Queen Formation on the Shattuck Escarpment, is massive quartzose siltstone/sandstone. This facies consists of gray (less commonly tan to red), recessively weathering, mature, well sorted coarse siltstone to fine sandstone. These strata are thickly bedded and generally lack all sedimentary structures. Minor bioturbation and haloturbation (in uppermost layers), can be seen. These sandstones typically define cycle bases and have a sheet-like geometry across the shelf, ranging from 1 to 30 ft in thickness, reaching a maximum thickness in the outer shelf and thinning in both updip and down-dip directions.

Cross laminated siltstone-sandstone.—

Cross-laminated siltstone-sandstone is texturally similar to the massive quartzose siltstone-sandstone. This facies is distinguished by the presence of ripple and small-scale low-angle trough cross stratification, and wavy current lamination. This facies is commonly found in the transitional zone where sandstone grades upward into carbonate within a cycle. In general these higher-energy sandstone facies are also better represented in more seaward portions of the shelf. Intervals of cross-laminated sandstone range from less than a foot up to 10 ft in thickness. An increase in percentage of carbonate peloids and cement is also observed as the sandstone changes upwards into carbonate facies.

Fusulinid-peloid packstone .—

Gray massive dolostone intervals with a faintly pelleted matrix and from 5 - 50 percent molds after fusulinid tests is a volumetrically small but stratigraphically critical facies in the Queen Formation. This facies may dominate more seaward portions of Grayburg platform cycles. Beds range in thickness from 1 ft to 30 ft, are laterally continuous, tapering in a landward direction. Rare large-scale seaward-dipping foreset stratification has been observed in one outcrop, but typically this facies is massive or shows signs of large vertical habitation burrows.

Fusulinid-peloid wackestone.—
This facies is similar to fusulinid-peloid packstone except that the matrix is dense, micritic, with low porosity and permeability and higher percentage of fusulinids preserved as molds rather than as preserved dolomitized tests.

Mollusc-algal-crinoid packstone-grain-dominated packstone.—
Poorly to moderately sorted bioclastic packstone containing fragments of molluscs, dasycladacean algae, and pelmatozoan debris in a peloidal matrix characterizes this facies. Rare cross stratification and vertical burrow tubes are the only sedimentary structures. Bedding ranges from 1 to 6 ft in thickness.

Peloid mudstone/wackestone.—
Dense dolomudstone with some pelleted fabric (wackestone) is present as 0.5 to 2 ft, and rarely 5 ft thick units. These mudstones are structureless and typically light gray in color. Because of its massive fine grained character, this facies typically has abundant small fractures and weathers recessively in outcrop. In the subsurface, these mudstones, which are low-energy facies, commonly have a higher proportion of clay minerals and organic debris and therefore have a characteristicly hotter gamma-ray signature.

In more basinward positions rare fusulinids or pelmatozoan debris may be found in these mudstones. In updip position mudstones and wackestones may be intercalated with fenestral dolowackestones and packstones. In this setting these fine-grained carbonates typically occur at or near the base of cycles.

Peloid packstone/grain-dominated packstone.—
Peloid dolopackstones and grain-dominated packstones are a common and somewhat generalized facies in the Grayburg and Queen Formations. This facies occurs as 1 to 3 ft thick units in all but the most seaward measured section. Peloid dolopackstones are grain-supported fabrics containing peloids (non-descript carbonate grains), ooids, bioclasts, or pisoids. This facies is typically structureless, but a few rare cross-bedded intervals have been included to distinguish them from the true ooid grainstones.

Ooid-peloid grainstone.—
Ooid-peloid grainstone is not an abundant facies in either the Grayburg or Queen Formations, but does form distinctive units up to 20 ft thick in the lower halved these formations. Ooids are up to 200 microns across and are well sorted. Small to medium scale trough and planar tabular cross stratification with some possible hummocky cross stratification are common. Some quartz sand is mixed with the ooids.

Fenestral/non-fenestral-pisolitic laminite.—
The fenestral-pisolitic laminite facies is used in a general sense here to include smooth to crinkly non-fenestral and fenestral cryptalgal laminites that may or may not display sheet cracks, pisolites and teepees. The common and widespread occurrence of these laminites, which typically define caps of upward shallowing cycles, make this facies one of the most useful for correlation and for monitoring changes in accommodation on the platform. These fenestral units also generally weather to resistant ledges making them easily traceable on photo mosaics. In subsurface these laminite intervals have distinctively low gamma signatures.

Depositional Model and Facies Tracts

A general depositional model based on observed facies relationships and cyclicity is critical in early stages of stratigraphic analysis. This depositional model is required as a tool to aid in decision-making when correlating unlike lithofacies within a chronostratigraphic framework. Simple procedures such as not correlating deeper-water facies down-dip into shallow-water facies within the same cycle are readily apparent examples of the use of depositional models. More complex interpretations are involved when evaluating the degree of depositional dip or paleotopography to use when correlating in a slope setting. Whereas the depositional model indicates using a 15 degree dip for correlation, most geologists is these situations abandon their conceptual models and fall back on lithostratigraphic correlations. Development of a depositional model that has a high-level of confidence is essential in trying to break the traditional mold of lithostratigraphic correlation.

The model presented here (Figs. 4 and 5) covers a broad platform to basin profile that draws to some extent on relationships from outside data sets we will examine. Major facies tracts that make up the depositional model are inner shelf, middle shelf, shelf crest, outer shelf, shelf margin, slope, toe of slope, and basin. Depositional environments described below are taken from times of high-frequency highstand deposition as these time slices show the full range of depositional environments. During cycle-scale lowstand sedimentation it is most likely that all but the most seaward portion of the outer shelf was emergent and covered by siliciclastic-dominated eolian ergs, reflecting little of the geomorphic variability that controls distribution of carbonate and mixed carbonate/siliciclastic facies tracts.

The model presented here is static and does not take into consideration the dynamically changing profile of the

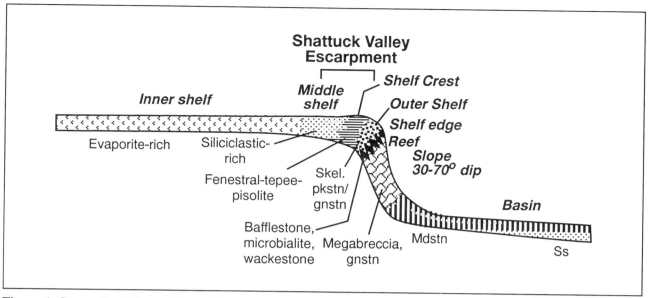

Figure 4. Generalized "static" depositional model for the Queen Formation showing relative position of facies tracts and their dominant facies.

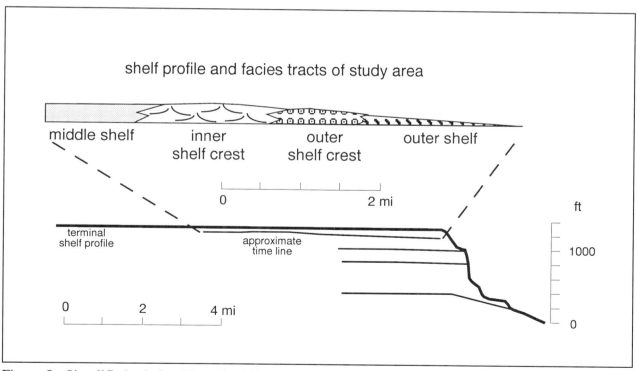

Figure 5. Simplified relationships of middle shelf through outer shelf facies tracts during Grayburg and Queen time. Facies tract widths are taken from the upper Grayburg Guadalupian 15 HFS.

platform during Grayburg or Queen sedimentation but rather averages out facies relationships in order to arrive at a set of terms for the facies tracts or environments that occurred during deposition of these units.

Inner Shelf.—

The inner shelf is that portion of the platform occupied by evaporitic flats and lagoonal deposits. The most likely depositional setting is that of a very shallow hypersaline lagoon with most of the evaporites being subtidally deposited sulfates. Tan to red siliciclastic mudstones and thin tongues of dolomudstone are also present in the inner shelf. A similar facies tract is documented by Sarg (1980) for the inner portions of the Seven Rivers shelf. We will not examine inner shelf deposits of the Queen Formation in outcrop but will get a chance to see some of these deposits in a core from the Keystone Field of the Central Basin Platform.

Middle Shelf.—

The middle shelf of the Queen Formation is a sandstone-dominated portion of the shelf situated between the inner shelf evaporitic lagoons and the shelf crest tract. The width of the middle shelf facies tract is estimated to be 2-4 mi in width, with the Shattuck Valley outcrops exposing only the most seaward third of that tract. The lack of exposure in this facies tract reflects recessive weathering of the sandstones, which are predominantly clay and carbonate-cemented. Most production from Queen Formation reservoirs is most likely from middle shelf settings, and it is believed that reservoir analogs may be exposed in the interior of the Guadalupes. A analogous depositional and reservoir setting in the younger Yates Formation is described by Borer and Harris (1991).

The sandstones that dominate this facies tract are primarily massive, making determination of depositional environment difficult. The paucity of higher energy depositional features such as cross bedding, cut and fill, or even small scale scour features favors a low-energy subtidal or peritidal setting somewhat protected from open-ocean environments. Some cycles in the middle shelf facies tract are capped by fenestrally laminated dolostones indicating that cycles shoaled to sea level. However, it is likely that numerous cycles are not clearly resolved in this setting because of amalgamation of individual sand units by bioturbation.

Shelf Crest.—

The shelf crest facies tract is critical in the platform profiles of the Guadalupian. This facies tract contains the highest energy deposits of the depositional profile, where fair-weather wave base impinges on the platform margin. Cycle stacking and the general distribution of this facies tract across the shelf through time is a useful indicator of longer term trends in accommodation on the platform. Lower Queen Formation shelf crest facies are sandstone, ooid grainstone, and fenestral pisolite. Amalgamated stacks of fenestral pisolitic facies up to 70 ft thick define this facies tract in the upper Queen Formation, forming massive cliffs. Water depths along the shelf crest were persistently the shallowest along the depositional profile. High energy grainstones and tidal flat facies disappear laterally as the shelf crest facies tract is traced into middle shelf and outer shelf settings. Water depths of 0 to 10 ft are likely.

In older ramp-style sequences of the San Andres Formation, the shelf crest (or it that case the ramp-crest) did not display a strong facies differentiation. An increase in percentage of ooid and peloid grainstone and minor increase in tidal flat facies is observed in these older sequences and some indication of a back-crest lagoon is demonstrated by the decrease in tidal flat facies symmetrically updip and down dip from the axis of the ramp crest. In Grayburg sequences, a somewhat more developed shelf crest is observed (Kerans and Nance, 1991) made up of inner ramp crest deposits where cycles are capped by thick fenestral tidal flat caps, and an outer shelf crest where cycles are dominated by thick ooid grainstone shoal complexes.

In the Queen Formation, the shelf crest becomes increasingly better developed through time, with lower cycles possessing an inner crest tidal flat capped cycle suite and an outer crest with less common grainstone shoal units. In the upper half of the Queen the shelf crest facies tract develops into the dominant component of the succession, comprising more than 50 percent of the total succession in the most seaward measured section. This trend towards increasingly well developed shelf crest facies tracts dominated by teepee-pisolite shoals persists from the Queen up into the Seven Rivers, Yates, and Tansill Formations. The shelf crest facies tract in the Queen Formation is a good example of cycle amalgamation, where a cliff forming unit of stacked fenestral dolostones can be traced paleolandward into the middle shelf setting and, in the approximate transition position, can be seen to consist of numerous discrete cycles.

Outer Shelf.—

The outer shelf facies tract is bounded updip by the tidal-flats and grainstone shoals of the shelf crest and down dip by shelf edge reefal deposits. This facies tract consists of 50% sandstone and 50% skeletal and peloidal dolopackstones. Fusulinid packstones and bioclastic packstones with pelmatozoan and dasycladacean fragments are most abundant in this tract. Cycles generally do not shoal up to peritidal conditions (rare fenestral capped cycles) and are generally thick (10-40 ft) and poorly defined.

The Queen outer shelf facies tract is closely comparable to that of the underlying Grayburg Formation, and also bears close resemblance to the outer shelf tract of the overlying Capitan-equivalent Seven Rivers, Yates and Tansill sequences. Although difficult to estimate accurately in this partially exposed succession, there probably existed between 20 and 60 ft of topographic relief difference across the outer shelf from the tidal flats and/or islands of the shelf crest to the shelf margin reefs. A slightly greater bathymetric change of 80-100 ft was suggested by Hurley (1989) for the Seven Rivers Formation outer shelf facies tract, otherwise referred to as fall-in beds.

Shelf Margin.—

The shelf margin of the Queen Formation is complicated by oversteepening and local collapse as documented by Franseen et al (1989; note that the Grayburg/Queen erosional scarp of Franseen et al is currently correlated by BEG studies to occur within the Queen Formation). One outcrop of the Goat Seep Formation, the Queen-equivalent shelf margin facies, is exposed in the extreme southern exposures of the Shattuck Valley. This exposure is similar to that described by Crawford (1981) from North McKittrick Canyon and the Western Escarpment, being a massive dolostone with traces of sponge and other metazoan fauna plus minor marine cement.

Slope.—

The slope facies tract of the Queen sequences, also included in the Goat Seep Formation, is exposed at the far southern end of the Shattuck Valley. The single 70 ft exposure is composed of dolomitized skeletal-intraclastic granulestone to grainstone with depositional dips of up to 50 degrees. Slope breccias along the Western Escarpment and in the North McKittrick Canyon area contain megabreccias up to 30 ft in thickness with clast sizes of several ft across. These deposits are characteristic of slope facies of reef-rimmed platforms.

Toe of Slope.—

The sedimentology of Queen-equivalent toe-of-slope deposits is poorly known. The South Wells Member of the Cherry Canyon Formation, and probably the upper portion of the Get Away member of the Cherry Canyon Formation, are Queen-equivalent toe-of-slope deposits. Getaway units are dominated by carbonate megabreccia but when traced farther from the shelf margin, are seen to include finer grained high-density sediment gravity flows and possible turbidites. A similar relationship can be shown for the South Wells member. Dips are significantly less than observed in the slope, between 1 and 10 degrees.

Basin.—

Basinal equivalents of the Queen-equivalent sequences are assumed to be virtually entirely sandstone and siltstone with only a few thin carbonate beds present.

The summary depositional model presented earlier for the Queen equivalent sequences, (Figs. 4 and 5) Guadalupian 16 and 17, is to-scale but contains significant generalization and averaging. As part of the course on high-frequency cyclicity we will attempt to enhance this model and characterize significant deviations from the model, attempting to understand these differences as reflecting changes in internal and/or external forcing functions such as the shift in position of accommodation during deposition. An example of one style of cycle development in the Queen Formation is shown in figure 6, which emphasizes the variations in cycle development across a broad platform due to differential preservation and diachroneity of a cycle. The descriptions given above serve only as a starting model, a model to be enhanced or discarded rather than one to which we will force-fit observations.

GEOLOGIC SETTING OF DATA SETS

Queen Formation of the Shattuck Escarpment

The Queen Formation data set is the focus of this course. These data are ideal for illustrating the use of tools of high-frequency sequence stratigraphy in construction of a high-resolution sequence framework. The excellent exposure of the outcrop from which the data were collected has permitted a high level of confidence of the correlations provided in the final answer sheet, allowing relatively unambiguous evaluation of interpretations carried out using the measured sections provided. Grouping of facies into cycles, cycles into cycle sets, and cycle sets into high-frequency sequences will be the main activity in this exercise. Correlation of cycle sets and high-frequency sequences between measured sections. Subsequent analysis of the Queen Formation is exposed over a large area of the central Guadalupe Mountains including southwestern Last Chance Canyon, Turkey Canyon, and Dark Canyon as well as in a continuous dip-oriented exposure along the Shattuck Escarpment (Fig. 7). The near-complete exposures along the eastern rim of the Shattuck Valley is the source of data for the Queen exercise (Fig 7, 8, 9).

The base of the exercise cross section is the Queen-Grayburg boundary of Hayes (1964) as taken from his maps and measured sections in the Shattuck Valley area. Paleokarst development along this Queen/Grayburg boundary in the Shattuck Valley area as well as additional stratigraphic data not presented here indicate that this boundary is a type 1 sequence boundary (Kerans and

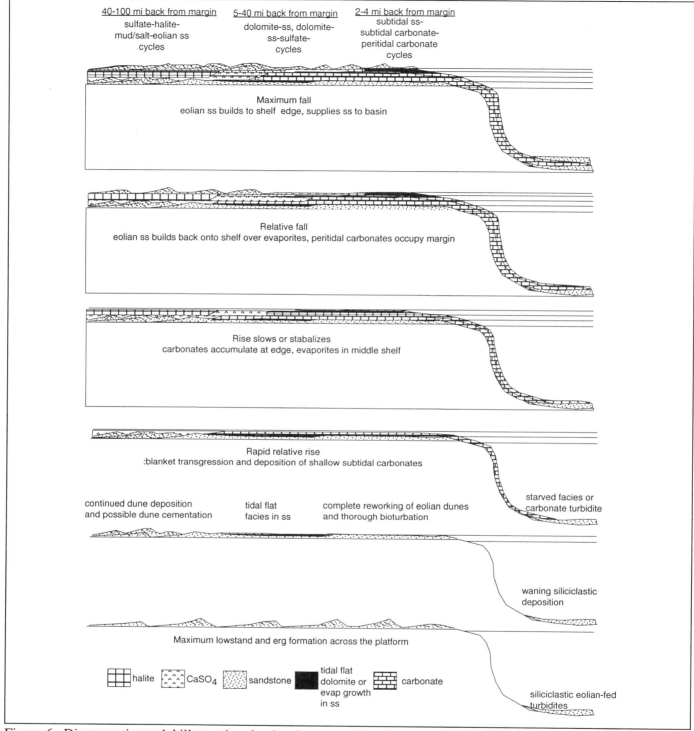

Figure 6. Diagramatic model illustrating the development of cycles in the Queen Formation. This model emphasizes the diachroneity of cycle formation even at the high-frequency scale. In this model, close the the shelf margin, cycle bases are picked at the base of thick marine-reworked sandstones (sandstone-subtidal carbonate-peritidal carbonate). In the inner portion of the middle shelf, eolian sandstones are commonly evaporite cemented and are not reworked so that the stratigraphic record of the basal flooding occurs at the base of the carbonate unit (carbonate-sandstone-evap plus sandstone cycle). In the most interior sections such as the Palo Duro Basin (Nance, 1988), eolian sandstones are partially to fully preserved and cycle base flooding occurs where gypsum deposits rest on eolian facies (sulfate-halite-mudsalt-eolian sandstone cycles).

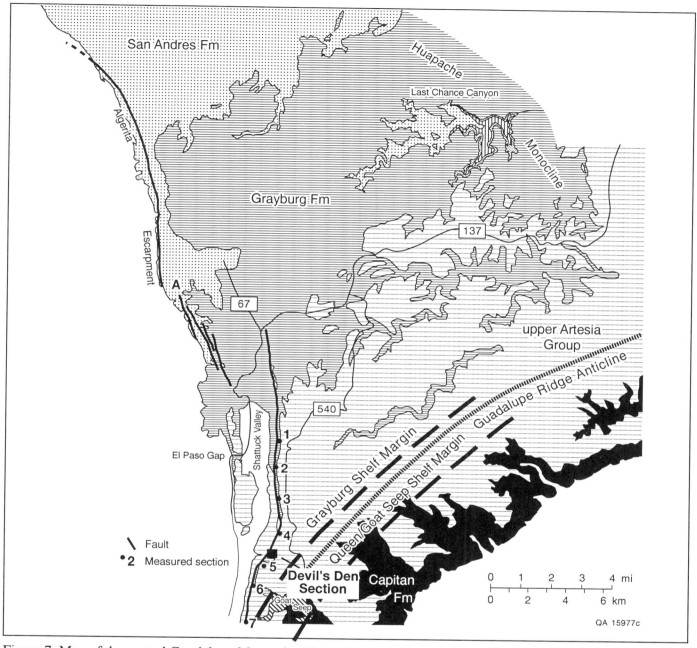

Figure 7. Map of the central Guadalupe Mountains illustrating the geologic setting of the Shattuck Valley area where the trip is focused. Measured sections that will be examined by teams, and that formed the basis of the existing stratigraphic framework of the Queen Formation are shown located. Map generalized from sources listed in figure 5.

Nance, 1991). Where complete, the Queen Formation is 350 ft thick. In the northern limit of the area of interest, the upper half of the Queen Formation is not exposed.

The upper 30-50 ft of the Queen Formation consists of the Shattuck member, a ripple to wavy-laminated siltstone/very fine sandstone that is widespread and continuous as a regional sandstone marker throughout the outcrop and the Northwest Shelf of the subsurface Permian Basin. This member is important in our exercise because it defines the top of the interval of interest and forms a useful datum. Whatever its exact depositional origin, the upper portion of the Shattuck member represents a classic example of facies-tract offset, as its upper portion of inner ramp origin shifts seaward to the shelf margin, stepping

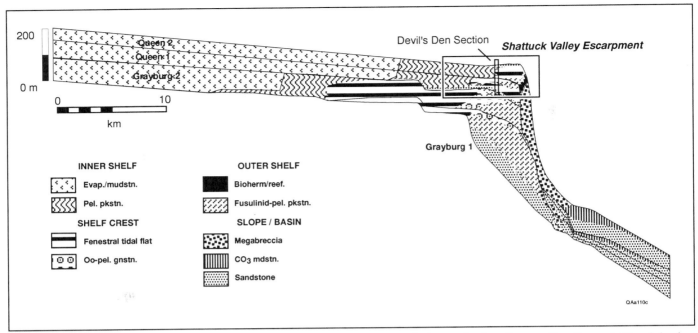

Figure 8. Stratigraphic framework of Grayburg and Queen sequences relative to the late Leonardian through Guadalupian stratigraphic framework. The Shattuck Escarpment are located in the stratigraphic framework.

out over shelf crest, outer shelf, and shelf margin facies tracts, and is almost continuous across the steep forereef slope and out into the basin.

Queen Formation shelf strata pass either conformably or unconformably into Goat Seep reefal and fore-reef facies and eventually into Cherry Canyon sandstone. Carbonate members of the Cherry Canyon Formation believed to be coeval with the Queen are the upper Getaway debris beds and the South Wells member (Fig 2).

The sections of the Queen Formation that we will work on in the correlation exercise are spaced out along the 3 mi of oblique dip exposure of the Shattuck Escarpment at approximately half-mile increments. More northerly sections of the Queen Formation are relatively thin and eventually become completely truncated at the northern end of the escarpment. This thinning reflects the northward increase in recessive-weathering siltstone and sandstone over more resistant carbonate strata as units are traced landward. The sections used as a framework for the cross section we will produce are located in figure 7. These sections include strata from middle shelf, shelf crest, and shelf-margin settings.

Grayburg Formation of the Texaco Holt Unit, North Cowden Field

The core and log cross sections from the Grayburg Formation at North Cowden field are useful for illustrating the character of high-frequency cycles and how these facies and cycles appear in subsurface data. The tie between core and wireline log is useful to demonstrate the log signature of the facies and cycles, and the correlation of the one cored well to several others along a dip line of cross section illustrates the added information derived from two-dimensional analysis of the cycles within the high-frequency sequence framework. These data also illustrate the level of resolution sought in order to constrain a detailed reservoir characterization.

The Grayburg Formation of the North Cowden reservoir is mixed siliciclastic/carbonate field with dominant production from siliciclastics that are concentrated at cycle bases in the lower of two high-frequency sequences. This reservoir is one of the largest Grayburg fields in the Permian Basin, and the only one along the eastern side of the Central Basin Platform where production is dominated by siliciclastic facies. It is situated along the eastern margin of the Central Basin Platform facing into the Midland Basin approximately 3 mi landward of the Grayburg shelf margin, resting almost directly on the terminal San Andres shelf margin (Figs. 1, 10). The core that will be viewed as part of this session comes from the most basinward, eastern side of the field and covers the lower 80 percent of the Grayburg Formation (Figs. 11, 12). We will examine both the stacking pattern of facies and cycles within the Grayburg core as well as those observed in two dimensions in a dip-oriented east-west section.

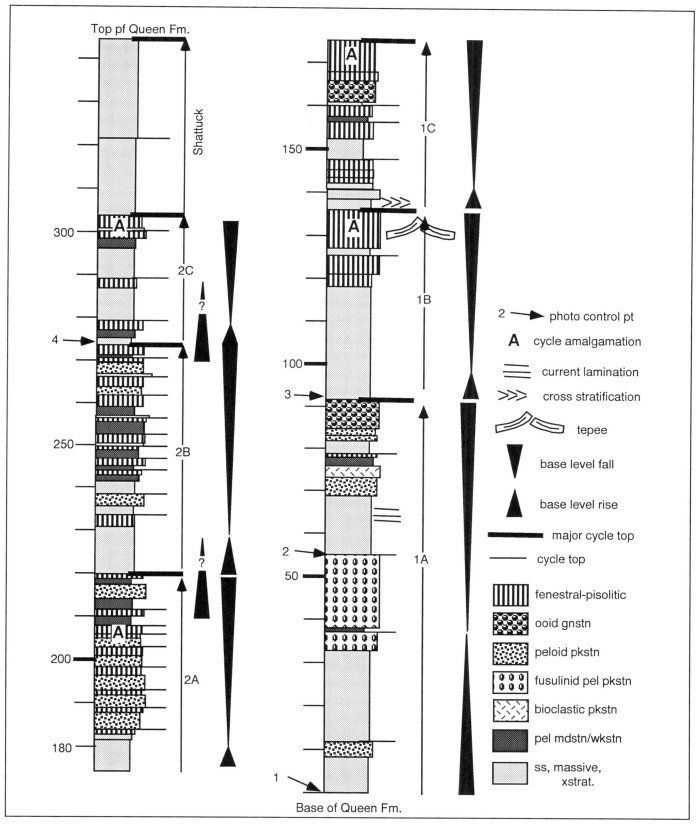

Figure 9. Measured section of the Queen Formation as exposed on the south wall of Devil's Den Canyon (see Fig. 7 for location). This section will serve as an introduction to interpreting the stratigraphic cyclicity within the Queen

USE OF ONE- AND TWO-DIMENSIONAL CYCLE ANALYSIS

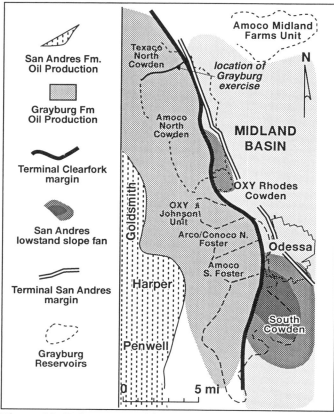

Figure 10. Paleogeographic map illustrating the structural and depositional setting of the North Cowden and related Grayburg reservoirs along the east side of the Central Basin Platform. The location of the line of section containing the Holt No. 75 core is shown at the northern end of the North Cowden field.

The Grayburg Formation at North Cowden displays the same two HFS recognized in outcrop along the Shattuck Valley Wall (Kerans et al. (1992) and along Plowman Ridge (Barnaby and Ward, 1995). The San Andres/Grayburg sequence boundary is a well developed karst surface. Outcrops in the Guadalupe Mountains display as much as 90 ft of vadose dissolution and collapse. In the subsurface, some 60 mi south of the North Cowden reservoir at Yates Field, paleokarst features are also well developed at the top-San Andres sequence boundary. The basal Grayburg HFS (Guadalupian 14 HFS) is characteristically rich in siliciclastic facies and displays a symmetrical stacking of transgressive and highstand cycles. In contrast, the upper Grayburg sequence contains a smaller proportion of siliciclastics and is dominated by aggradationally stacked cycles. A distinctive fusulinid-bearing set of cycles within the transgressive systems tract commonly extends well onto the shallow-water platform at the base of the Guadalupian 15 HFS.

QUEEN EXERCISE

1. Discuss facies development, interpretation of cycles, and cycle hierarchies as seen in the section at Devils Den (Fig. 9) as a group. Identify cycles, cycle sets, and candidates for longer-term cycles. Discuss cycle stacking trends, as well as trends in facies proportions, possible facies tract offset, and other pertinent elements of the section. Identify potential key correlation surfaces.

2. Divide into teams and work individual Queen sections along the dip-cross section as previously carried out

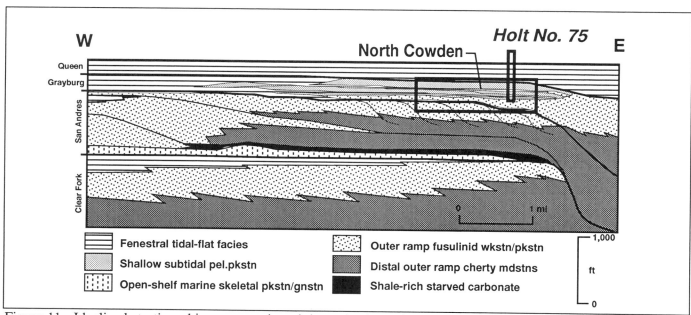

Figure 11. Idealized stratigraphic cross section giving seismic-scale image of setting of Grayburg strata within the North Cowden reservoir. Relative position of the Holt No. 75 core is given.

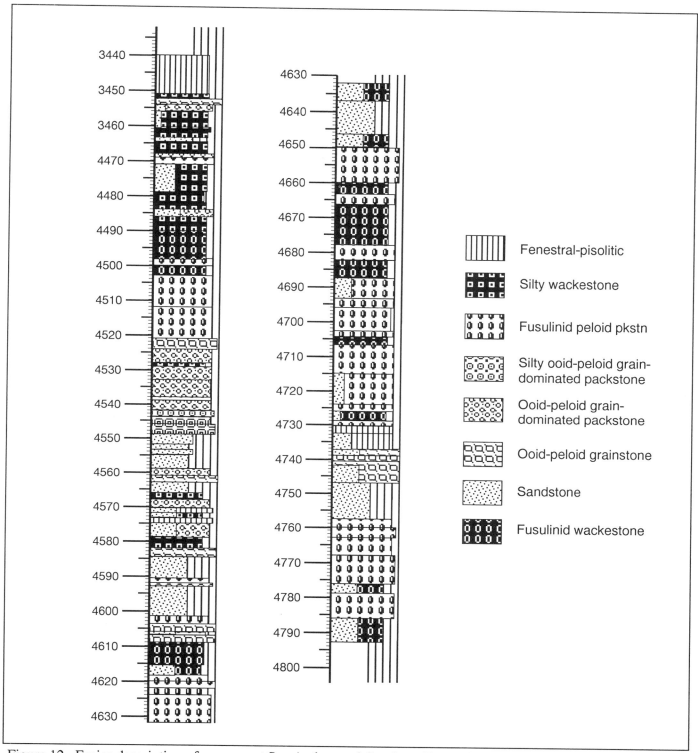

Figure 12. Facies description of uppermost San Andres and Grayburg Formation intervals in the Holt No. 75 core.

by group as a whole. Each group will present the results of their analyses.

3. Carry analysis of Queen strata into two dimensions using uninterpreted dip-oriented cross section from Shattuck Valley Wall. The procedure is as follows

Construct high-frequency sequence framework.

a. Systematically interpret cycles and cycle sets on each measured section one at a time *before* correlating sections.

b. Examine individual interpreted sections and evaluate whether potentially correlative cycle sets are distinguishable.

c. Correlate cycle set boundaries across the section as best as possible.

d. Within each cycle set, attempt to correlate individual cycles between sections.

Analyze individual cycle sets and sequences within framework

e. Identify facies tracts within cycle sets and sequences by means of isolating correlatable cycles with common vertical facies successions. It is critical at this stage that *key facies* are identified. Key facies are those that have a well constrained and understood depositional origin, such as cross-stratified ooid grainstone, or fenestral algal laminite. Another aspect required of key facies is that they be common across multiple cycle sets or sequences, allowing comparison of facies and facies tract occurrence across multiple chronostratigraphic units.

f. Using facies tracts delineated by identification of key facies within cycles, establish a base-level curve and link this to the identification of high-frequency sequences within the cross section. Discuss changing facies tract attributes in terms of evolution of controling parameters.

GRAYBURG EXERCISE

The Grayburg exercise will follow the same steps as outlined for the Queen Formation, with the addition of two initial steps.

1. A core of the Texaco Holt No. 75 will be present (Fig. 12). We will work together in the time remaining to point out facies, cycles, and their recognition on wireline logs.

2. We will then move to the uninterpreted cross-section of the Texaco Holt unit and analyze this in the fashion discussed previously for the Queen exercise, emphasizing integration of wireline log data.

SUMMARY

This overview of the application of high-frequency sequence stratigraphy to one and two-dimensional data sets illustrates the application of tools of high-frequency sequence analysis to both one and two-dimensional data. The use of these techniques to develop detailed stratigraphic frameworks suitable for refined reservoir characterization in a central message of this approach. Recognition of a nested hierarchical set of cycles, including high-frequency cycles, high-frequency sequences, and composite sequences, is outlined and enforced. The analysis of strata using cycle-stacking patterns, facies proportions, cycle symmetry, stratal preservation, and facies tract offset to first construct a chronostratigraphically significant framework, and subsequently to analyze these strata in terms of external forcing functions linked to changing accommodation, sediment supply, or climate will allow for improved predictive analysis.

ACKNOWLEDGMENTS

This course was developed as part of the Bureau of Economic Geology's Carbonate Reservoir Characterization Research Laboratory. Funding for this project was from industry sponsors Agip, Amoco, ARCO, British Petroleum, Chevron, Conoco, Exxon USA and Exxon Production Research, Fina, JNOC, Marathon, Mobil, Phillips, Shell, Texaco, Total, and Unocal. The Department of Energy supplied matching funds for portions of this research under ANNEX I to DE-FG22-89BC14403, and DE-AC22-89BC14470. Drafting was by the Bureau of Economic Geology, University of Texas at Austin. This manuscript has benefited from review by S. Nance. Page layout is by Margret Evans.

REFERENCES

BARNABY, R. J., AND WARD, W. B, 1995, Sequence stratigraphic framework, high-frequency cyclicity and three-dimensional heterogeneity: Grayburg formation, Brokeoff Mountains, New Mexico. *in* Pause', P.H., and Candelaria, M.P., (eds.), Carbonate Facies and Sequence Stratigraphy: Practical Applications of Carbonate Models: Permian Basin Section- Society of Economic Paleontologists Mineralogists Publication 95-36, p. 37.

BORER, J. M. AND HARRIS, P. M., 1991, Depositional facies and cyclicity in the Yates Formation, Permian Basin - Implications for reservoir heterogeneity: American Association of Petroleum Geologists Bulletin, v. 75, p. 726-779.

CRAWFORD, G. A., 1981, Depositional history and diagenesis of the Goat Seep Dolomite (Permian, Guadalupian), Guadalupe Mountains, West Texas-New Mexico: unpublished Ph.D. dissertation, University of Wisconsin, Madison, Wisconsin, 300 p.

FRANSEEN, E. K., FEKETTE, T. E., AND PRAY, L. C., 1989, Evolution and destruction of a carbonate bank at the shelf margin: Grayburg Formation (Permian), western escarpment, Guadalupe Mountains, Texas, in Crevello, P. D., Wilson, J. L., Sarg, J. F., and Read, J. F., eds., Controls on carbonate platform and basin development: Society of Economic Paleontologists and Mineralogists Special Publication Number 44, p. 289-304.

GARDNER, M. H., 1992, Sequence stratigraphy of eolian-derived turbidites: deep water sedimentation patterns along an arid carbonate platform and their impact on hydrocarbon recovery in Delaware Mountain Group reservoirs, West Texas, in D. H. Mruk and B. C. Curran, eds., Permian Basin Exploration and Production Strategies: Applications of Sequence Stratigraphic and Reservoir Characterization Concepts, West Texas Geological Society, Inc. Symposium, Publication Number 92-91, p. 7–11.

GARDNER, M.H., BARTON, M.D., TYLER, N., AND FISHER, R.S., 1992, Architecture and Permeability Structure of Fluvial-Deltaic Sandstones, Ferron Sandstone, East-Central Utah, in Flores, R.M., (ed) Mesozoic of the Western Interior, Society of Economic Paleontologists and Mineralogists Field Trip Guidebook, p.5-20.

GOLDHAMMER, R. K., OSWALD, E. J. AND DUNN, P. A., 1991, Hierarchy of stratigraphic forcing: Example from Middle Pennsylvanian shelf carbonates of the Paradox Basin, in Franseen E. K., Watney, W. L., Kendall, C. G. St. C. and Ross W., eds., Sedimentary Modeling: Computer Simulations and Methods for Improved Parameter Definition: Kansas Geological Survey Bulletin 233, p. 361-413.

HANFORD, C.R. AND LOUCKS, R.G., 1994, Carbonate depositional sequences and systems tracts-responses of carbonate platforms to relative sea-level changes in: Loucks, R.G., and Sarg, J.F., (eds) Carbonate Sequence Stratigraphy recent developments and applications, American Association of Petroleum Geologists, Memoir 57, p 3-41.

HAYES, P. T., 1964, Geology of the Guadalupe Mountains, New Mexico: United States Geological Survey Prof. Paper 446, 69 p.

HURLEY, N. F., 1978, Facies mosaic of the lower Seven Rivers Formation (Permian), North McKittrick Canyon, Guadalupe Mountains, New Mexico: Unpublished M.S. thesis, University of Wisconsin, Madison, Wisconsin, 198 p.

HURLEY, N. F., 1989, Facies mosaic of the lower Seven Rivers Formation, McKittrick Canyon, New Mexico, in Harris, P. M., and Grover. G. A., (eds.), Subsurface and outcrop examination of the Capitan Shelf Margin, northern Delaware Basin Society of Economic Paleontologists Mineralogists Core Workshop Number 13, p. 325-346.

JAMES, N. P., 1979, Shallowing-upward sequences in carbonates, in Walker, R. G., ed., Facies Models: Geoscience Canada Reprint Series 1, Toronto, p. 109-119.

KERANS, C., W.M. FITCHEN, M.H. GARDNER, M.D. SONNENFELD, S.W. TINKER, AND B.R. WARDLAW, 1992, Styles of sequence development within uppermost Leonardian through Guadalupian strata of the Guadalupe Mountains, Texas and New Mexico; in D.H. Mruk and B.C. Curran, eds., Permian Basin Exploration and Production Strategies: Applications of Sequence Stratigraphic and Reservoir Characterization Concepts, West Texas Geological Society, Inc. Symposium, Publication Number. 92-91, p. 1–7.

KERANS, C., LUCIA, F.J., AND SENGER, R.K., 1994, Integrated characterization of carbonate ramp reservoirs using Permian San Andres Formation outcrop analogs. Bulletin American Association of Petroleum Geologists, v. 78, number 2, p. 181-216.

KERANS, CHARLES, LUCIA, F. J., SENGER, R. K., FOGG, G. E., NANCE, H. S., KASAP, EKREM, AND HOVORKA, S. D., 1991, Characterization of reservoir heterogeneity in carbonate ramp systems, San Andres-Grayburg Formations, Permian Basin: The University of Texas at Austin, Bureau of Economic Geology, Reservoir Characterization Research Laboratory Final Report, 245 p.

KERANS, C. AND NANCE, S., 1991, High-frequency cyclicity and regional depositional patterns of the Grayburg Formation, Guadalupe Mountains, New Mexico, in Meader-Roberts, Sally, Candelaria, M. P., and Moore G. E., (eds.), Sequence stratigraphy, facies and reservoir geometries of the San Andres, Grayburg, and Queen Formations, Guadalupe Mountains, New Mexico and Texas: Permian Basin Section-Society of Economic Paleontologists Mineralogists Publication 91-32, p. 53-69.

KING, P. B., 1948, Geology of the southern Guadalupe Mountains: United States Geological Survey Professional Paper 215, 183 p.

KOERSHNER, W.F., AND J.F. READ, 1989, Field and modeling studies of Cambrian carbonate cycles, Virginia Appalachians, Journal of Sedimentary Petrology, v. 59, p. 654–687.

MEISSNER, F. F., 1970, Cyclic sedimentation in Middle Permian strata of the Permian Basin, West Texas and New Mexico: West Texas Geological Society publication 72-60, p. 118-142.

MITCHUM, R. M., JR., AND J. C. VAN WAGONER, 1991, High-frequency sequences and their stacking patterns: sequence-stratigraphic evidence of high-frequency eustatic cycles, in: K.T. Biddle and W. Schlager, eds., The Record of Sea-Level Fluctuations: Sedimentary Geology, v. 70, p. 131–160.

MORAN, W. R., 1954, Proposed type sections for the Queen and Grayburg Formations of Guadalupe age in the Guadalupe Mountains, Eddy County, New Mexico: (abst.) GSA Bull., vol. 65, p. 1288, and Guidebook of Southeastern New Mexico, New Mexico Geological Society p. 147-150.

NANCE, H. S., 1988, Interfingering of evaporites and red beds: An example from the Queen/Grayburg formation, Texas: in Kocurek, G., ed., Late Paleozoic and Mesozoic eolian deposits of the western interior of the United States, Sedimentary Geology, v. 56, p. 357-382.

PARSLEY, M. J., 1988, Deposition and diagenesis of a late Guadalupian barrier-island complex from the middle and upper Tansill Formation (Permian), east Dark Canyon, Guadalupe Mountains, New Mexico: unpublished M.A. thesis, The University of Texas at Austin, Austin, Texas, 247 p.

READ, J. F., GROTZINGER, J. P., BONE, J. A. AND KOERSCHNER, W. F., 1986, Models for generation of carbonate cycles: Geology, v. 14, p. 107-110.

READ, J. F., AND GOLDHAMMER, R. K., 1988, Use of Fischer plots to define 3rd order sea-level curves in peritidal cyclic carbonates, Ordovician, Appalachians: Geology, v. 16, p. 895-899.

RUPPEL, S. C., 1992, Controls of platform development in the Leonard Series (Middle Permian) of West Texas: Significance of multifrequency cyclicity and paleotopography: American Association of Petroleum Geology Annual convention, abstract, p. 112.

SARG, J. F., 1980, Petrology of the carbonate-evaporite facies transition of the Seven Rivers Formation (Guadalupian, Permian), southeast New Mexico: Journal of Sedimentary Petrology, v. 50, p. 73-96.

SARG, J. F., AND LEHMANN, P. J., 1986, Lower-middle Guadalupian facies and stratigraphy, San Andres/Grayburg Formations, Guadalupe Mountains, New Mexico, in Moore, G. E., and Wilde, G. L., (eds.), San Andres/Grayburg Formations: Lower-middle Guadalupian facies, stratigraphy, and reservoir geometries, Guadalupe Mountains, New Mexico: Permian Basin Section-Society of Economic Paleontologists and Mineralogists, Publication. Number 86-25, p. 1-8.

SARG, J. F., ROSSEN, C., LEHMANN, P. J., AND PRAY, L. C., (eds.), 1988, Geologic guide to the Western Escarpment, Guadalupe Mountains, Texas: Permian Basin Section, Society of Economic Paleontologists and Mineralogists, Publication 88-30, 60 p.

SONNENFELD, M.D., AND CROSS, T. A., (1994) Volumetric partitioning and facies differentiation within the Permian upper San Andres Formation of Last Chance Canyon, Guadalupe Mountains, New Mexico; in Loucks, R.G. and Sarg, J. F., eds., Recent advances and applications of carbonate sequence stratigraphy: American Association of Petroleum Geologists Memoir .

TYRRELL, W. W., JR., 1969, Criteria useful in interpreting environments of unlike but time-equivalent carbonate units (Tansill-Capitan-Lamar), Capitan Reef Complex, West Texas and New Mexico, in Friedman, G. M., (ed.), Depositional environments in carbonate rocks: Society of Economic Paleontologists and Mineralogists Special Publication 14, p. 80-97.

VAIL, P. R., 1987, Seismic stratigraphy interpretation procedure, in A.W. Bally, ed., Atlas of Seismic Stratigraphy, volume 1, American Association of Petroleum Geologists Studies in Geology no. 27, p. 1–10.

VAIL, P. R., MITCHUM R.M. AND THOMPSON, S. III, 1977, Seismic stratigraphy and global changes of sea level, Part 3: relative changes of sea level from coastal onlap; in Payton, C.W., ed., Seismic stratigraphic application to hydrocarbon exploration: American Association of Petroleum Geology, Memoir 26, p. 63-97.

VAN WAGONER, J. C., POSAMENTIER, H. W., MITCHUM, R. M., JR., VAIL, P. R., SARG. J. F., LOUTIT, T. S. AND HARDENBOL, J., 1988, An overview of the fundamentals of sequence stratigraphy and key definitions, in Wilgus, C. K., Hastings, B. S., Kendall, C, G. St. C., Posamentier, H. W., Ross, C. A. and Van Wagoner, J. C., eds., Sea-Level changes: An Integrated Approach: Society of Economic Paleontologists and Mineralogists Special Publication Number 42, p. 39-46.

VAN WAGONER, J. C., R. M. MITCHUM, K. M. CAMPION, AND V. D. RAHMANIAN, 1990, Siliciclastic sequence stratigraphy in well logs, cores and outcrops: concepts for high–resolution correlation of time and facies: American Association of Petroleum Geology, Methods in Exploration Series, number 7, 55 p.

WANLESS, H. R., 1991, Observational foundation for sequence modeling, in Franseen E. K., Watney, W. L., Kendall, C. G. St. C. and Ross W., eds., Sedimentary Modeling: Computer Simulations and Methods for Improved Parameter Definition: Kansas Geological Survey Bulletin 233, p. 43-62.

PART 3

SEQUENCE STRATIGRAPHY AND RESERVOIR DELINEATION OF THE MIDDLE PENNSYLVANIAN (DESMOINESIAN), PARADOX BASIN AND ANETH FIELD, SOUTHWESTERN USA

by L. James Weber, J.F. (Rick) Sarg and Frank M. Wright

SEQUENCE STRATIGRAPHY AND RESERVOIR DELINEATION OF THE MIDDLE PENNSYLVANIAN (DESMOINESIAN), PARADOX BASIN AND ANETH FIELD, SOUTHWESTERN USA

L. JAMES WEBER,[1] J. F. (RICK) SARG,[1] AND FRANK M. WRIGHT[2]

[1]Mobil Exploration and Producing Technical Center, P. O. Box 650232, Dallas, Texas 75265-0232 and
[2]Mobil Exploration and Producing Technical Center, P. O. Box 819047, Farmers Branch, Texas 75381-9047

ABSTRACT

A sequence stratigraphic framework has been established for the Middle Pennsylvanian (Desmoinesian) section in southeastern Utah using: 1) surface exposures at Honaker Trail, Raplee Anticline, and Eight Foot Rapids located 25 to 40 miles (40-64 km) west of the Aneth field, 2) core and well logs in SE Utah, SW Colorado, NW New Mexico, and NE Arizona, and 3) regional seismic. Bounding discontinuities (sequence boundaries and maximum flooding surfaces) have been correlated over several thousand square miles in the Four Corners region. Systems tracts of 3rd-order composite sequences (0.5-5.0 m.y.) are comprised of 4th-order sequences (0.1-0.5 m.y.) and 5th-order depositional cycles or parasequences (0.01-0.1 m.y.). Nineteen discrete and mappable high-frequency depositional cycles are recognized within three fourth-order depositional sequences of the Desert Creek and lower Ismay section (Middle Desmoinesian) at the McElmo Creek Unit of the Aneth field, southeastern Utah. These simple sequences stack into parts of two third-order sequence sets.

Facies analysis of 12,000 feet (3,660 m) of core was tied into the chronostratigraphic framework to constrain correlation of high-frequency depositional cycles (parasequences). Facies mapping within parasequences permitted 1) prediction of porous and permeable facies and 2) characterization of variability in reservoir pore systems. Syndepositional dolomitization and dissolution in peritidal facies, at shoal crests of parasequences and at sequence boundaries caused modification of reservoir character.

Since discovery of the Aneth field (1956), 370 million barrels of oil (~1.3 billion barrels original oil in place) have been produced from Middle Pennsylvanian carbonates of the Ismay and Desert Creek intervals. Stratified reservoirs occur within lowstand, transgressive, and highstand systems tracts. Siltstone, dolostone, and evaporites form lowstand wedges that were deposited 150 feet below the crest of the Aneth carbonate platform. Porous dolomudstone and dolowackestone are productive where they onlap and pinch out against the Aneth carbonate platform and are isolated from reservoirs on the platform. Within transgressive systems tracts, lagoonal/tidal flat dolomudstone/wackestone compose parasequences and display intercrystalline and solution-enhanced secondary porosity. Core analysis and production/performance data indicate that significant fluid pathways are developed in dolomudstone deposited on the carbonate platform on paleodepositional highs. In the lower Desert Creek, initial parasequences of the highstand systems tract represent a time of mound building and platform development as a result of coalescing biologic communities of phylloid algae. Interparticle and shelter porosity dominate. Subsequent parasequences within the lower Desert Creek highstand systems tract are composed of skeletal and nonskeletal wackestone to grainstone. Porosity is developed on paleodepositional highs at the top of parasequences where shoal water facies have preserved primary pore systems that are secondarily enhanced by leaching of less stable carbonate minerals by meteoric water. Reservoirs dominated by primary pore systems provide the best long term production and account for the majority of oil produced in McElmo Creek. In the upper Desert Creek highstand systems tract, ooid/peloid grainstones aggrade and prograde to fill available depositional space. Hydrocarbons are produced on the platform and along the platform to basin margin from carbonate sand sheets and allochthonous debris aprons. Grainstone debris aprons may also be deposited during early lowstand conditions of the lower Ismay sequence. On the platform meteoric diagenesis resulted in the formation of oomoldic porosity in ooid grainstone deposits beneath the upper Desert Creek sequence boundary. Within moldic pore systems storage capacity is favorable, but permeability is low, generally less than 1 md. Facies composed dominantly of moldic porosity in the absence of significant primary porosity are poor reservoirs.

INTRODUCTION

Upper Paleozoic rocks in the Paradox basin of southwestern Colorado and southeastern Utah contain large hydrocarbon resources. Deeply incised canyons along the San Juan River provide spectacular exposures of rocks lateral to the oil producing reservoirs of nearby hydrocarbon fields. These outcrops are utilized along with selected cores, well logs, and seismic to examine the mixed carbonate-siliciclastic-evaporite depositional system that may be analogous to other oil producing basins.

Particular attention is devoted to deposition on the carbonate shelf and Aneth platform of the Paradox basin. Evaporite cycles are observed in the central basin and are correlated to the shelf and platform where evaporites are absent. The application of sequence stratigraphy has proved very helpful in the recognition, interpretation, and correlation of these cycles and depositional sequences.

Sequence stratigraphic analysis has allowed an improved characterization of the stratified Middle Pennsylvanian carbonate reservoirs of the McElmo Creek Unit in the Greater Aneth Field. A regional stratigraphic framework establishes major chronostratigraphic surfaces (e.g., sequence boundaries and flooding surfaces) that constrain correlation of high-frequency depositional cycles. In the Aneth field, correlation of genetically-related depositional cycles forms the basis of the stratigraphic layer model. The distribution of porous and permeable facies is used to predict the size and continuity of reservoirs. Production/performance (engineering) data are integrated along with the geology to test, refine, and validate the geologic model and to develop an understanding of interwell- and reservoir-scale heterogeneities that affect fluid flow and hydrocarbon recovery. This approach has led to 1) a better understanding of platform to basin correlations and associated field performance anomalies, 2) a better understanding of interwell heterogeneity related to changes in depositional facies, 3) unit-wide mapping of time equivalent geologic layers, 4) prediction of reservoir quality by relating deposition and early diagenesis, and 5) a first-pass evaluation of flow zones. Paleodepositional topography, relative changes in sea level, sediment accumulation, climate, and early diagenesis each influence facies development and reservoir quality at the McElmo Creek Unit.

The major objectives of this paper are: 1) to show how outcrop and subsurface data are combined to develop a reservoir model with a sequence stratigraphic framework, 2) to describe a stratigraphic architecture at McElmo Creek that comprises genetically-related mappable units, and 3) to use the architecture to describe the distribution, continuity, and quality of select reservoir facies at McElmo Creek. Methods and concepts employed here are applicable to exploration and producing ventures worldwide.

GEOLOGIC BACKGROUND

The Paradox basin is a late Paleozoic basin of Pennsylvanian and Early Permian age which is located in southeastern Utah and southwestern Colorado entirely within the Colorado Plateau physiographic province (Figure 1). Oil productive Middle Pennsylvanian limestone, dolostone, and sandstone are available for study at several outcrop localities and within the gross pay section of the Greater Aneth field. Measured sections of exposed Pennsylvanian rocks are located approximately 40 miles (64 km) west of the Aneth oil field in southeastern Utah (Figure 2). The Pennsylvanian section at Honaker Trail and at Aneth is located along the southwest margin of the Paradox basin in an area refered to as "Canyon Lands" (Figure 3).

Oil is produced along the southwestern margin of the Paradox basin in Middle Pennsylvanian cyclic carbonates. The Greater Aneth Complex is the largest oil field in the Paradox basin (Figure 4). The Aneth field has an estimated original oil-in-place of 1.3+ billion barrels and an esti-

Figure 1. Map showing general location of Paradox basin (after Herrod and others, 1985).

mated ultimate recovery of approximately 600 million barrels. Cummulative oil production is approaching 370 million barrels (1/1/93). The producing reservoirs are highly stratified. Porosity is developed in several facies. Algal banks in the Desert Creek and Ismay intervals of the Paradox Formation (Figure 5) are cited in the literature as the dominant reservoir facies (see Choquette, 1983 for review). Average drilling depth is about 5,750 feet (1,754 m), and the wells are on 20 to 80 acre spacing.

McElmo Creek Unit is entirely within the Navajo Indian Reservation and is situated along the eastern portion of the Greater Aneth Field (Figure 6). Within McElmo Creek, oil is produced primarily from the Desert Creek and to a lesser extent within the overlying lower Ismay. Oil productive facies are stratigraphically controlled to the north, east, and south by porosity and permeability pinchouts. Porous facies continue to the west, but an oil-water contact at -960 feet (-293 m) subsea limits production in that direction.

The term "platform" is used to describe paleodepositional high areas (i.e., the Greater Aneth Platform or the Akah platform in the vicinity of Honaker Trail). The Aneth Platform is an isolated carbonate platform that is approximately 75 square miles (200 sq. km.) in areal extent. Platform development is controlled largely by colonization and growth of phylloid algae in the lower Desert Creek. Thinning of the lower Desert Creek away from the Greater Aneth field area results from the absence of algal banks. Beyond the Aneth Platform, the lower Desert Creek is generally less than 10 feet (3 m) thick. The depositional low surrounding the Aneth Platform is referred to as the "basin". On the platform, the lower Desert Creek may exceed 100 feet (30 m) in thickness. An intervening slope exists between the platform and basin. Interwell calculations suggest a minimum slope angle of 5 degrees, however, 30 to 40 degrees probably approximates the norm, especially during early Desert Creek time, when the biologic system was responsible for platform development.

LITHOSTRATIGRAPHY

The general stratigraphic nomenclature of upper Paleozoic and Mesozoic rocks in southeastern Utah is shown

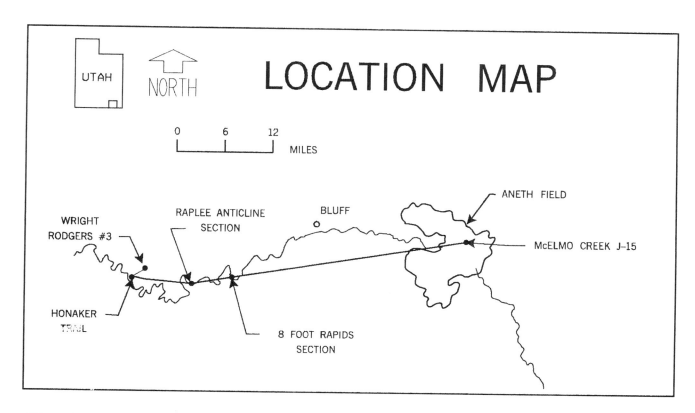

Figure 2. Location map showing Aneth Oil Field and various outcrop locations. Line segments connecting outcrop locations and well in Aneth Field indicate traverse for cross-section that is shown later.

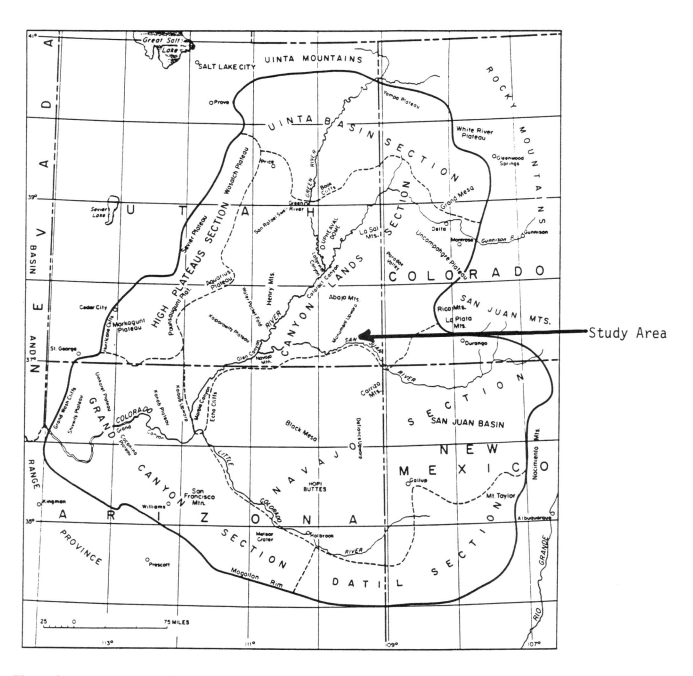

Figure 3. Index map of Colorado Plateau province with sectional boundaries (after Hunt, C.B., U.S. Geological Survey Professional Paper 279; and Thornbury, 1965).

in Figure 7. Stratigraphic terminology follows the work of Wengerd and Strickland (1954) and Baars (1962). Lower and middle Paleozoic strata exceed 2,100 feet (640 m) in thickness and rest on Precambrian basement. Cambrian, Devonian, and Mississippian carbonates and siliciclastics are not exposed in the study region. However, Pennsylvanian and Permian systems are well-exposed along the San Juan River and are encountered in the subsurface within and adjacent to the Aneth field. Pennsylvanian rocks range from 1,000 feet (300 m) in northeastern Arizona to 2,000 feet (600 m) in the Aneth area, and exceed 5,000 feet (1,500 m) in the depocenter adjacent to the Uncompahgre Uplift (Figure 8). Paleozoic rocks are overlain by a thick Mesozoic section, but Laramide uplift and erosion has removed a considerable portion of this younger section (Molenaar, 1981). The Phanerozoic stratigraphy in the Aneth area is comprised of Cambrian through Jurassic rocks (Figure 5).

The geology of the Middle to Late Pennsylvanian Hermosa Group is the primary focus of this paper. The Hermosa Group has a gradational contact with the underlying Molas Formation and is divisible into the Pinkerton Trail, Paradox and Honaker Trail formations in ascending order. The Molas Formation (see Figure 7) overlies a paleokarst surface developed on the Mississippian Redwall and Leadville limestones and spans the Mississippian-Pennsylvanian boundary. The Honaker Trail Formation is overlain by the Permian Cutler Group (Wengerd, 1973).

The reader is referred to the following authors for detailed descriptions of the local and regional stratigraphy:

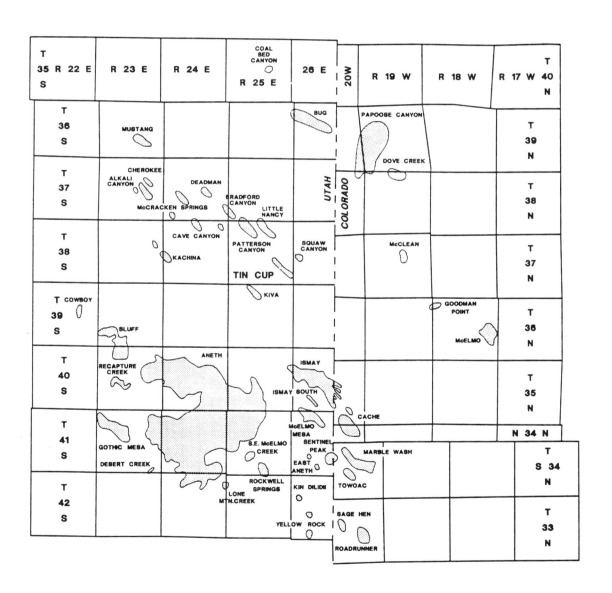

Figure 4. Map showing productive fields in Paradox basin. Largest field is the Greater Aneth field (after Herrod and Gardner, 1988).

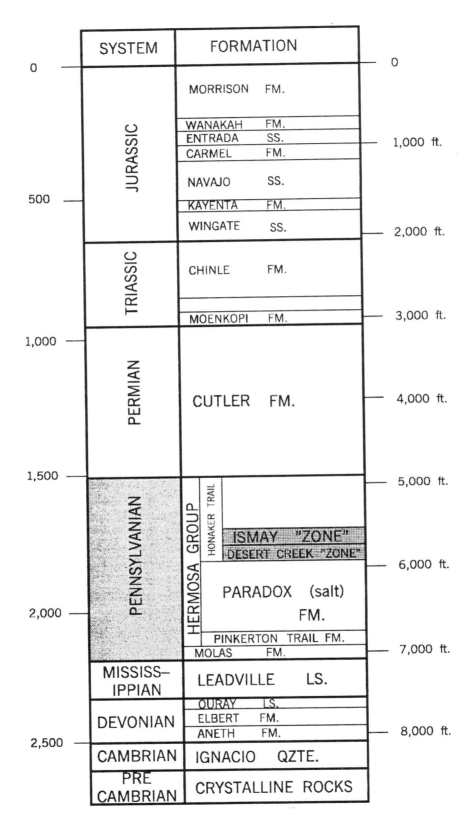

Figure 5. Stratigraphic column with associated drilling depths in feet and meters, Aneth field area (after Peterson, 1992).

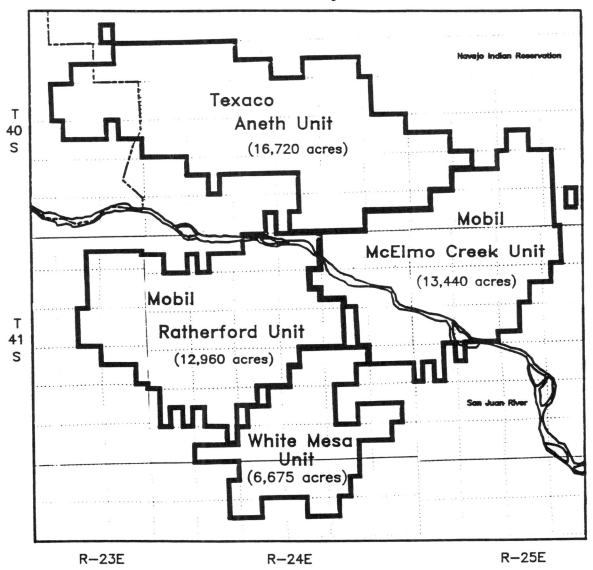

Figure 6. Map showing the location of the four operational units of the Greater Aneth field. The focus of this study is on the McElmo Creek Unit.

Clair (1952), Wengerd and Strickland (1954), Herman and Sharps (1956), Mallory (1958), Wengerd and Matheny (1958), Elston and others (1962), O'Sullivan (1965), Peterson and Hite (1969), Molenaar (1981), Doelling (1983), Mack and Rasmussen (1984), and Peterson (1992).

MOLAS FORMATION

The Molas Formation rests unconformably on Mississippian carbonates and is composed of red to purple siltstone, sandstone, and shale with associated thin limestone beds (Wengerd and Matheny, 1958; Merrill and Winar, 1958). The basal portion of the Molas contains limestone rubble and boulders from the underlying Mississippian Redwall and Leadville formations. The Molas Formation varies considerably in thickness (1-150 feet). It was deposited as the sea transgressed the craton and as a result, is a diachronous unit that ranges in age from Late Mississippian (Chesterian) to late Early Pennsylvanian (earliest Desmoinesian) (Wengerd and Matheny, 1958). Overlying the Molas Formation is the Pinkerton Trail Formation. The contact is gradational and is placed above the uppermost fine-grained red siltstone or shale bed, beneath the first gray shale and limestone interval of the Pinkerton Trail Formation (Wengerd and Matheny, 1958).

PINKERTON TRAIL FORMATION

Like the Molas Formation, the Pinkerton Trail Formation is time-transgressive, ranging in age from Early to Middle Pennsylvanian (Atokan-Desmoinesian) (Wengerd and Matheny, 1958). The Pinkerton Trail Formation is not exposed in the study area, but core has been recovered through the interval. The Pinkerston Trail Formation is composed of fossiliferous crinoidal and fusulinid-rich limestone with light to dark gray shale, siltstone, and dolostone. According to Wengerd and Strickland (1954), the Pinkerton Trail Formation is relatively thin and ranges in thickness from 0 to 200 feet (0-60 meters).

PARADOX FORMATION AND RELATED ROCKS

The Paradox Formation is defined formally at the base and top by the first and last occurrence of evaporites. Evaporites are interbedded with limestone, dolostone, shale, and coarser grained siliciclastics. In areas where evaporites are not present, only the Pinkerton Trail and Honaker Trail formations are recognized in the Hermosa Group (Baars, Parker, and Chronic, 1967).

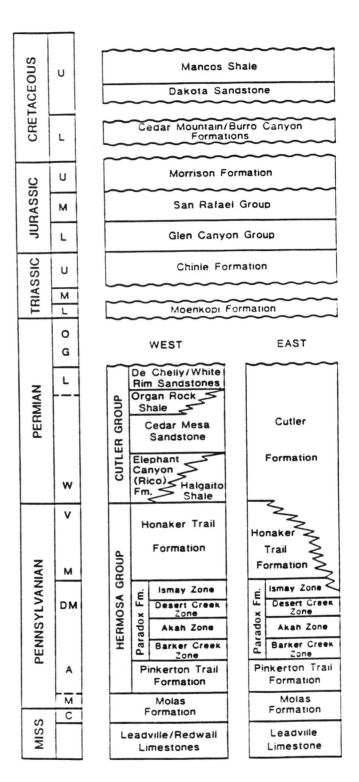

Figure 7. Generalized upper Paleozoic and Mesozoic stratigraphy in the Paradox basin region. Terminology for Pennsylvanian and Permian systems adapted from Wengerd and Strickland (1954) and Baars (1962).

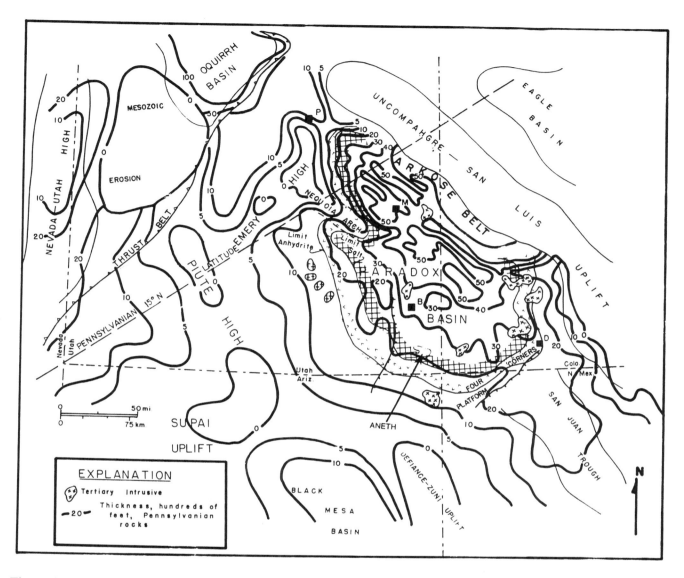

Figure 8. Thickness map, Pennsylvanian System of the Paradox basin and surrounding area. Edge of Paradox basin evaporites and basinal and uplifted areas are shown. Arkose belt on east side of basin grades westward into evaporites and organic rich carbonate facies (taken from Peterson, 1992).

The Paradox Formation and Related Rocks are subdivided into three broad areally distributed facies zones. Proximal to the Uncompaghre Uplift and extending to the south and west to the central portion of the Paradox basin is a thick clastic wedge of quartz arenite and arkose. In the central basin area siliciclastics interfinger with thick evaporite units that are punctuated by thin organic-rich dolostone and shale. Estimates of original depositional thickness of bedded halite, anhydrite, dolostone, and shale range from approximately 4,000 to 7,000 feet (1,220-2,135 m) (Hite, 1960; Peterson and Hite, 1969). Laterally equivalent to the clastic and evaporite units in the southern and southwestern portion of the Paradox basin is the carbonate shelf facies (Wengerd and Strickland, 1954; Wengerd and Matheny, 1958; Peterson and Hite, 1969). According to Wengerd and Matheny (1958), the shelf facies of the Paradox Formation and Related Rocks are as much as 1,000 feet (305 m) thick. However, near local source areas (e.g., Monument upwarp, Zuni-Defiance Uplift, etc.), the carbonate shelf facies thins drastically and is siliciclastic-rich. The exposed carbonate facies at the Honaker Trail section consist of fossiliferous limestone and barren dolostone interbedded with calcareous fine-grained sandstone, siltstone, and shale. Similar rock types are observed along with evaporites at the Raplee Anticline section and in the subsurface at the Aneth Field. In addition, conspicuous carbonate mounded buildups are observed in this part of the Paradox basin. They are composed of phylloid algae and form important petroleum reservoirs.

In the evaporite portion of the basin, the Paradox Formation is divided into four chronostratigraphic intervals that are traceable. These time-rock units are, from oldest to youngest, the Barker Creek, Akah, Desert Creek, and Ismay stages (Figure 7). A fifth time-rock unit, the Alkali Gulch, is recognized in the central part of the basin and underlies the Barker Creek. More detailed descriptions of carbonate rocks in the Paradox Formation are given by Wengerd (1951, 1955), Choquette and Traut (1963), Peterson and Hite (1969), Hite (1970), Wilson (1975), and Choquette (1983).

In the Greater Aneth Field area, the Desert Creek carbonate section extends from the base of the underlying black laminated dolomudstone, informally named the Chimney Rock, to the base of the overlying black laminated dolomudstone, informally called the Gothic. Above the Desert Creek are lower Ismay carbonates which are capped by the Hovenweep, another black laminated dolomudstone. A typical 1950's GNT neutron and Gamma Ray log is used to show the stratigraphy through the gross reservoir section (Figure 9).

HONAKER TRAIL FORMATION

The Honaker Trail Formation overlies the Paradox Formation (in the absence of evaporites, the Honaker Trail Formation overlies the Pinkerton Trail Formation) and reaches a thickness of 1,750 to 1,900 feet (534-580 m) (Fetzner, 1960; Wengerd and Matheny, 1958). Fossiliferous limestone is interbedded with sandstone, siltstone, and shale and grades into arkosic clastics of the undifferentiated Cutler Formation to the northeast toward the Uncompahgre Uplift. Overlying the Honaker Trail Formation are thin beds of red fine-grained sandstone, siltstone, and shale with subordinate gray sandstone and marine limestone.

CUTLER GROUP AND CUTLER FORMATION UNDIFFERENTIATED

The Cutler Group consists of marine and nonmarine sedimentary rocks. In the eastern portion of the basin along the flank of the Uncompahgre Uplift, the undifferen-

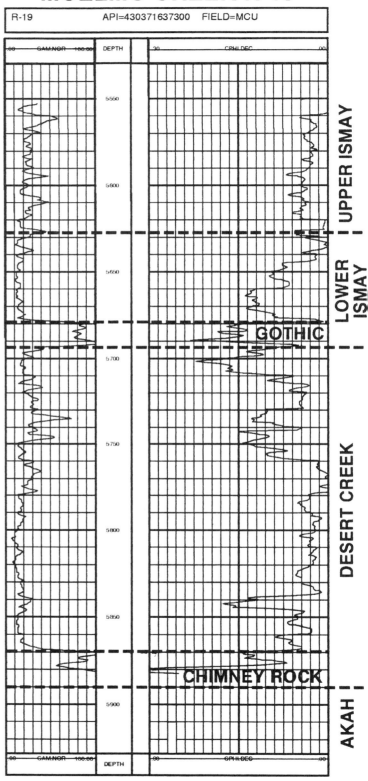

Figure 9. Type log that shows lithostratigraphy of McElmo Creek Unit of the Aneth field.

tiated Cutler Formation exceeds 15,000 feet (4,575 m) (Baars, 1979) and consists dominantly of arkosic sandstone. The lowermost part of the Cutler Formation interfingers with evaporites and carbonates of the Paradox and Honaker Trail formations. Mallory (1958), Elston and others (1962), Mallory (1972), Campbell (1979, 1980) and Mack and Rasmussen (1984) ascribe alluvial fan, fluvial, and eolian environments for the Cutler Formation.

On the southern and western portion of the Paradox basin, the Cutler Group is divided into several Wolfcampian and Leonardian age formations (see Figure 4) (Baars, 1979). In this area, the Honaker Trail Formation is overlain either by the fluvial-plain to marginal-marine Halgaito Shale (Baars, 1962, 1979) or the laterally equivalent Elephant Canyon Formation, a sequence of interbedded mudstone, siltstone, sandstone, marine limestone and anhydrite that exceeds 1,000 feet (305 m) in thickness at the type section (Baars, 1962). The Halgaito Shale and Elephant Canyon Formation are conformably overlain by the Cedar Mesa Sandstone, an extensive, well-sorted quartzarenite with prominent large-scale cross-stratification. In the area of Honaker Trail, the Cedar Mesa Sandstone reaches 1,400 feet (427 m) and has been interpreted as marginal-marine and eolian (Baars, 1962, 1979; Mack, 1977, 1979), and as exclusively eolian in origin (Loope, 1984). The Organ Rock Shale overlies the Cedar Mesa Sandstone and consists of red to reddish brown, fine-grained clastic rocks that are similar to the Halgaito Shale and probably were deposited on a lowland fluvial plain adjacent to nearshore marine environments (Baars, 1962, 1979). The DeChelly Sandstone, a reddish orange sandstone with large-scale high-angle cross-bedding is eolian in origin and overlies the Organ Rock Shale. In general, rocks of the Cutler Group coarsen to the north and east, and become more arkosic as they grade laterally into the undifferentiated Cutler Formation.

PALEOGEOGRAPHIC SETTING

The Paradox basin is an elongate northwest- to southeast-oriented depression of Late Paleozoic age (Figure 10). During the Middle Pennsylvanian, the Paradox basin was an enclosed basin approximately 200 miles (320 km) long and 100 miles (160 km) wide. Tectonic activity in the Paradox basin was at a peak during the Middle Pennsylvanian (Atokan and early Desmoinesian), resulting in major uplift of the Uncompahgre region, which shed clastic debris into the northern and eastern portion of the subsiding Paradox Basin (Baars, 1966). The deepest part of the asymmetrical basin is adjacent to the Uncompahgre Uplift where a thick succession (5,000 to 8,000 feet; 1,525 to 2,440 m) of alternating black, organic-rich shale and dolostone, evaporites, and quartz clastics were deposited. The basin shallows as a series of half grabens step up onto the more tectonically stable western and southwestern carbonate shelf (Figure 11) (Baars and Stevenson, 1982). Here, the periodically emergent shelf was marked by repetitive shoaling cycles of shallow marine carbonates, restricted marine carbonates and shales, and evaporites.

Paleoreconstructions based on paleoclimatic and paleomagnetic data suggest that the Paradox basin was located approximately 15 to 20 degrees north latitude of the paleoequator during the Middle Pennsylvanian (Heckel, 1977 and 1980) (Figure 12). If this interpretation is correct, the Paradox basin would have been located in a subtropical, semi-arid to arid climatic belt. Parrish (1982) and Heckel (1977 and 1980) postulate that the prevailing wind direction was from present day north to south.

Surrounding the Paradox basin were a number of positive highlands or source areas which were either active contributors of sediment, mostly quartz arenites or arkoses or were periodically submerged positive elements with significantly thinned Pennsylvanian rock sections. The Paradox basin was bounded on the east and northeast by the Uncompaghre/San Luis uplifts, on the northwest by the Emery and Circle Cliffs Uplift, on the southwest by the Kaibab Uplift, on the south by the Zuni-Defiance Uplift, and on the southeast by the Four Corners Platform (Figures

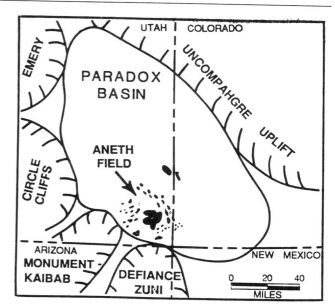

Figure 10. Paradox basin bounded by major uplifts (after Herrod and others, 1985).

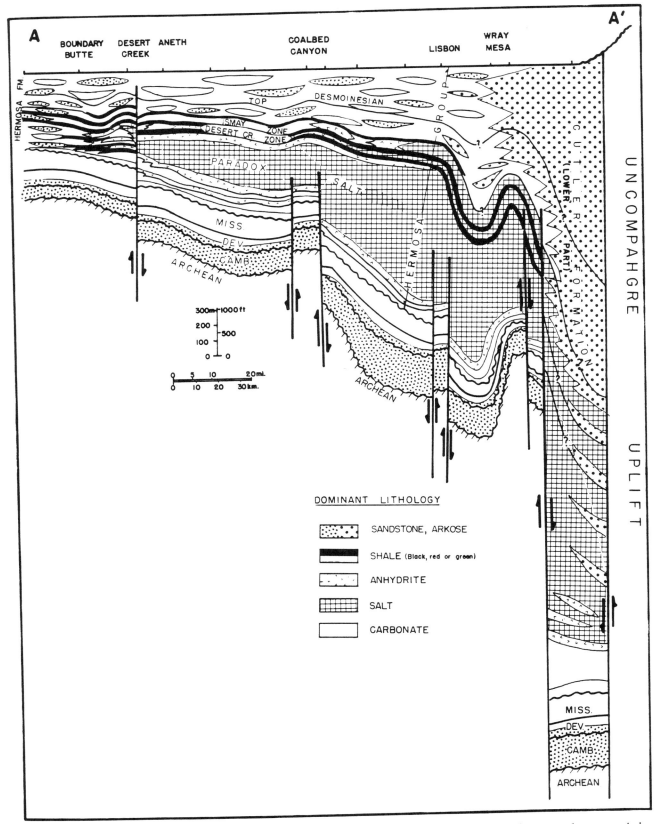

Figure 11. North-south paleostructural and lithologic cross-section A-A' that extends from northeastern Arizona to the Uncompahgre Uplift. Datum is the top of the Pennsylvanian System (taken from Peterson, 1992).

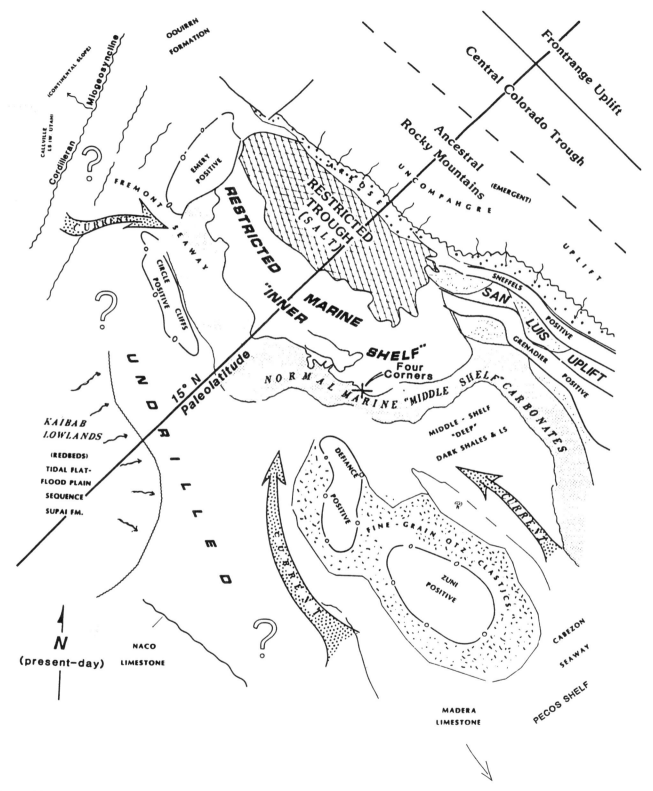

Figure 12. Paleogeography of the Greater four corners area at Late Middle Desmoinesian time (Desert Creek) showing the distribution of major depositional facies in the Paradox basin.

12 and 13). Four accessways led into and out of the Paradox basin. Of these inlets, the Oquirrh and Fremont embayments were unlikely contributors of open-marine water (Hite, 1970; Elias, 1963; Fetzner, 1960; Wengerd and Matheny, 1958). The Oquirrh embayment is a structural sag between the Uncompaghre and Emery uplifts, and the Fremont embayment is a structural sag between the Emery and Kaibab uplifts. A third accessway was postulated by Hite (1970) to be present between the Kaibab and Zuni-Defiance uplifts in Arizona. However, the Kaibab Uplift extended from the Defiance Uplift to the Emery Uplift (Blakey, 1980; Szabo and Wengerd, 1975, Fetzner, 1960), making the south to southwest edge of the Paradox basin completely landlocked.

The fourth and most probable opening, the Cabezon accessway was located along the southeast margin of the Paradox basin, between the Zuni-Defiance Uplift and the Uncompaghre Uplift (Hite, 1970; Ohlen and McIntyre, 1965; Wengerd and Matheny, 1958; Herman and Sharps, 1956; Fetzner, 1960). Here, water flowed over the Four Corners Platform, which underwent minor tectonic uplift during the Pennsylvanian (Peterson and others, 1965), resulting in shallow-water conditions on the platform. Fetzner (1960) and Peterson and others (1965) postulated that a submarine clastic fan was shed off the Uncompaghre Uplift, further restricting this accessway.

TECTONO-STRATIGRAPHIC HISTORY

Few drill holes penetrate Precambrian basement in the Four Corners area (Four Corners refers to the place where the borders of Arizona, New Mexico, Colorado, and Utah

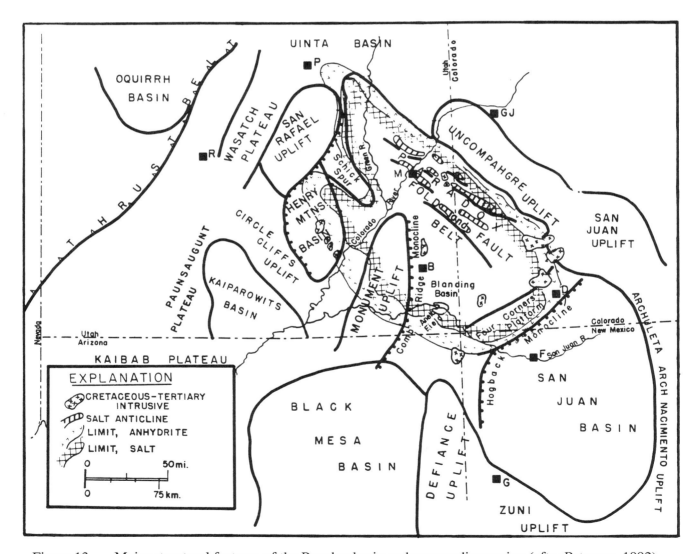

Figure 13. Major structural features of the Paradox basin and surrounding region (after Peterson, 1992).

meet). Several nearby areas have experienced uplift and expose intensely deformed Precambrian basement composed of igneous and metamorphic rocks. Baars and Stevenson (1981) have studied exposed Precambrian rocks in the San Juan Mountains, and coupling this data with gravity and magnetic surveys and seismic data, they delineate a series of northwest trending basement rifts which are cut by subordinate northeast trending fractures. They postulate that vertical movement throughout the Paleozoic created shoaling conditions which controlled the formation of Devonian offshore carbonate sand and sandstone bars, Mississippian crinoidal banks, and Middle Pennsylvanian algal bioherms. These carbonate accumulations have been exploration targets in the Paradox basin for 40 or more years.

From Cambrian to Mississippian time, the Four Corners region was part of the foreland shelf of the Cordilleran miogeosyncline, a stable shelf that accumulated shallow water carbonates and clastic material. The area experienced regional erosion and/or non-deposition during the Ordovician, Silurian, and Early Devonian. During the Middle Mississippian, limestone was deposited on an areally extensive shallow carbonate platform which extended from Arizona to Alberta, Canada (Figure 14; see Figure 7 for general stratigraphic column).

Collision of the South American plate with North America began in the Late Mississippian (Merimac-Chester) and resulted in overthrusting along the Marathon-Ouachita suture zone on the southern margin of North America (Kluth and Coney, 1981). This pulse of collision culminated in late Chester time with the break-up of the southwestern region of the U.S. into a series of uplifted basement blocks flanked by subsiding basin areas (e.g., Central Basin Platform, Permian Basin, Uncompaghre Uplift, Southwestern Colorado) (Figure 15).

An accompanying major eustatic fall in sea level resulted in subaerial exposure of the Mississippian carbonate platform over the western North America region (Saunders and Ramsbottom, 1986). Subaerial exposure continued through the Early Pennsylvanian and the nonmarine Molas Formation siliciclastics were deposited on the eroded and karsted Mississippian limestone.

A second episode of thrusting culminated in uplift in the late Atokan and initiated a period of rapid subsidence in the Paradox basin. Marine transgression occurred and the fossiliferous limestone and shale of the Pinkerton Trail Formation were deposited in the center of the basin. Along the periphery of the basin, dolomitic shales and siltstones were deposited. With continued, rapid subsidence of the area, salt deposition expanded and reached a maximum during the Alkali Gulch, Barker Creek, and Akah cycles of the Pennsylvanian Paradox Formation. Evaporite accumulation became more restricted in areal extent during the Desert Creek, Ismay, and Honaker Trail cycles (Berghorn and Reid, 1981). Evaporite accumulation exceeded 5,000 feet (1,525 m) in the center of the basin during the time represented by the Middle Pennsylvanian. Episodic uplift of the ancestral Uncompaghre basement block occurred in the Pennsylvanian, and 15,000 to 20,000 feet (4,575-6,100 m) of relief may have existed between the crestal portion of the Uncompaghre Uplift and the Paradox basin during the Late Pennsylvanian (Frahme and Vaughn, 1983). A thick wedge of arkosic sandstone (the Honaker Trail Formation) gradually filled the Paradox basin. Shallow marine sands and fluvial sediments of the Cutler Group further infilled the basin during the Lower Permian.

As the Paradox basin subsided during Pennsylvanian time, sea level oscillated as a result of glaciation on the Gondwana supercontinent. Meter-scale asymmetric cycles were deposited in response to these relative sea level changes. As many as 35 to 40 cycles have been identified through the Desmoinesian section in the Paradox basin. Duration of cycles was variable and probably ranged from 10's to 100's of thousands of years. Ultimately, the cyclicity that is observed is probably related to one or more of the following driving mechanisms: 1) eustatic sea level changes as a result of glaciation, 2) changes in basin subsidence and tectonism, and 3) changes in rate and type of clastic influx (Peterson, 1992). Meter-scale cycles stack into larger cycles that are correlative interregionally during the Pennsylvanian System (Ross and Ross, 1985 and this paper).

Relative changes in sea level caused repeated flooding and exposure of the carbonate shelf. Black calcareous "shales" were deposited during transgressive events; carbonates and evaporites accumulated during sea level highstands and lowstands, respectively (Figure 16). Depending on water depth, salinity, and distance from siliciclastic influx, conditions were favorable locally for phylloid algae to flourish during highstands of sea level. During several cycles, phylloid algae grew into extensive algal mound complexes. These mounds grew to wave base and coalesced laterally in relatively shallow water. Algal mound facies are identified in only a few cycles. Desert Creek and Ismay mounds are productive at the Greater Aneth field.

Beginning in Early Permian time, rapid subsidence of the basin ceased and marine waters withdrew to the west

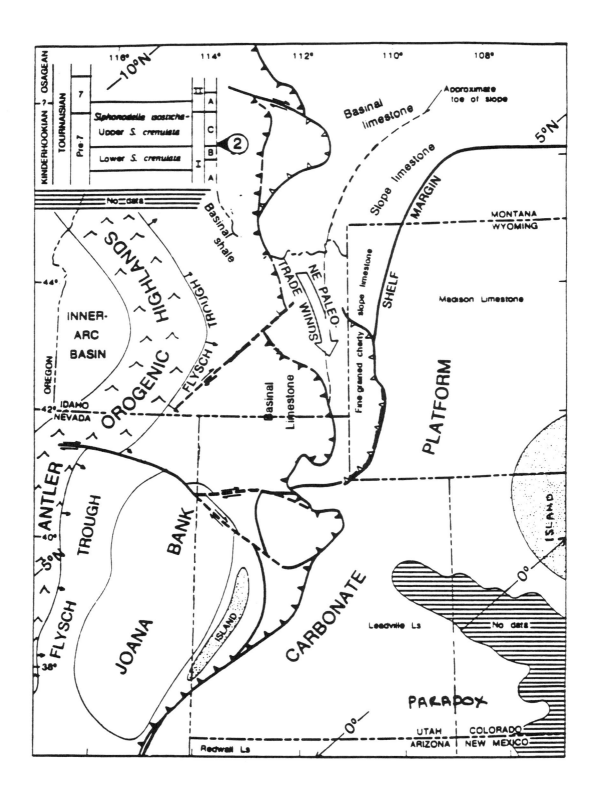

Figure 14. Paleogeographic map: earliest well-defined carbonate platform during deposition of Mississippian Leadville Limestone (after Gutschick, Sandberg, and Sando, 1980).

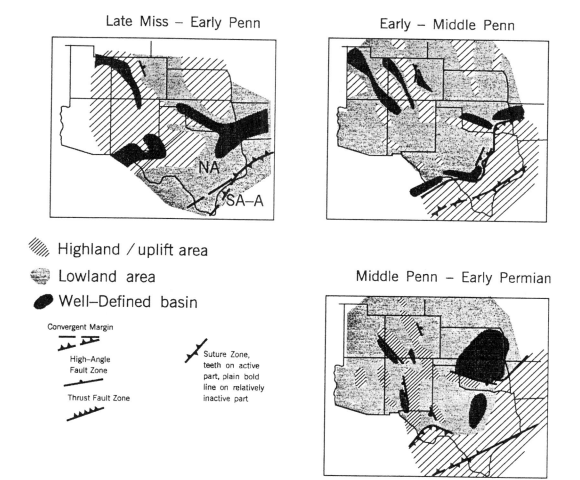

Figure 15. Time-slice paleotectonic reconstructions, southwestern U.S.

as continental shales and sands spread across the basin. The Uncompahgre and San Luis uplifts underwent rejuvenation and were eroded as indicated by arkosic alluvial fans of the Cutler Formation adjacent to these uplifts. Sediment load near the Uncompahgre appears to have created sufficient compressional force to initiate salt flowage along the northwest trending salt anticline belt.

In the Paradox basin, salt movement continued periodically into the Jurassic when major growth appears to have ceased. During Cretaceous time, the basin became a part of the Rocky Mountain geosyncline and was bordered on the west by uplifted and thrusted terrain in response to the Sevier Orogeny. Over 6,000 feet (1,830 m) of marine and floodplain sediments were deposited in the geosyncline. During latest Cretaceous and early Tertiary time, the Laramide Orogeny occurred. Middle to Late Tertiary faulting, collapse and graben formation took place in the salt anticline area, probably due to canyon cutting, dissolution, and salt flowage. Igneous activity also occurred in many parts of the basin forming stocks and laccolithic domes. Regional uplift, volcanism, and erosion from Late Tertiary to the Present are responsible for the topography of the area.

FACIES DESCRIPTIONS AND INTERPRETATIONS

Middle Pennsylvanian (Desmoinesian) rocks were investigated in the San Juan River Canyon along the Goosenecks area of southeastern Utah. With the exception of the Pinkerton Trail Formation and the lower Barker Creek, the remainder of the Desmoinesian succession is exposed in the river canyon. Parts of this section have been measured and described by Pray and Wray (1963), Wengerd (1973), Hite and Buckner (1981), Goldhammer and others (1991), and Gianniny (in prep.). The stratigraphic sections at Honaker Trail, Raplee Anticline, and Eight Foot Rapids were measured and described to supplement preexisting work (Figures 17, 18, 19, oversized, see note at back). Laterally continuous rock faces were observed over several miles and contribute toward a greater understanding of lateral lithologic variability.

Seventy-nine cores representing more than 12,000 feet (3,660 m) of Akah, Desert Creek, and Ismay section were examined. Geologic interpretations of depositional environments and diagenesis were refined through detailed petrographic examination of approximately 500 thin-sections. Fourteen major facies are identified. The facies described here are similar in many respects to facies described from subsurface core and rock exposures in other parts of the Paradox Basin (Choquette and Traut, 1963; Pray and Wray, 1963; Elias, 1963; Peterson and Ohlen, 1963; Gray, 1967; Hite, 1970; Hite and Buckner, 1981; Baars and Stevenson, 1981; and Goldhammer and others, 1991). Salient features of facies are displayed in tabular format (Table 1). Facies descriptions presented here are based primarily on subsurface rock descriptions within the Greater Aneth Field area. Descriptive observations and facies interpretations of subsurface rock data are essentially identical, and thus, directly applicable to Desmoinesian-aged rocks exposed along the San Juan River Canyon. Similar facies are recognized both in outcrop and subsurface in the southern and southwestern portion of the Paradox basin.

ANHYDRITE LITHOFACIES

The anhydrite lithofacies is areally restricted to the basin and attains greatest thickness, 30 feet (10 m), in topographically low areas. The anhydrite facies onlaps and pinches out against the carbonate platform. This facies is dominated by white to light gray anhydrite exhibiting a variety of primary to displacive/replacive textures. Bedding textures are massive with palmate structures (Plate 1-A) [all Plates are at the end of the text], less commonly mosaic/nodular with chickenwire fabric or thin-laminated with fine-grained dolomite. Subvertical to oblique anhydrite columns several centimeters tall with irregular stylolitic margins are separated from one another by brown, laminated, microcrystalline dolomudstone (Plate 1-A). Mosaic anhydrite exhibits nodules that are white, spherical to elliptical, and 3 to 50 mm in diameter. Mosaic anhydrite is found in association with thinly laminated anhydrite interlayered with brown-gray dolomud-stone/wackestone. Dolomite and anhydrite occur as mm thick, wavy, crinkled sheets.

The anhydrite lithofacies was deposited dominantly in a shallow hypersaline subaqueous environment. Periodic desiccation of salinas resulted in subaerial exposure and possible short-term sabkha conditions. Subaqueous deposition is supported by the occurrence of local anhydrite thicks in topographic low areas in the basin. In addition, the volumetrically important massive anhydrite with palmate growth structures is consistent with shallow subaqueous evaporite deposition (Warren and Kendall, 1985). Subvertical to oblique columns reveal crystal orientation that is normal to depositional bedding (Plate 1-B). Water depth is estimated to range from 0 to 33 feet (0-10 m) in the

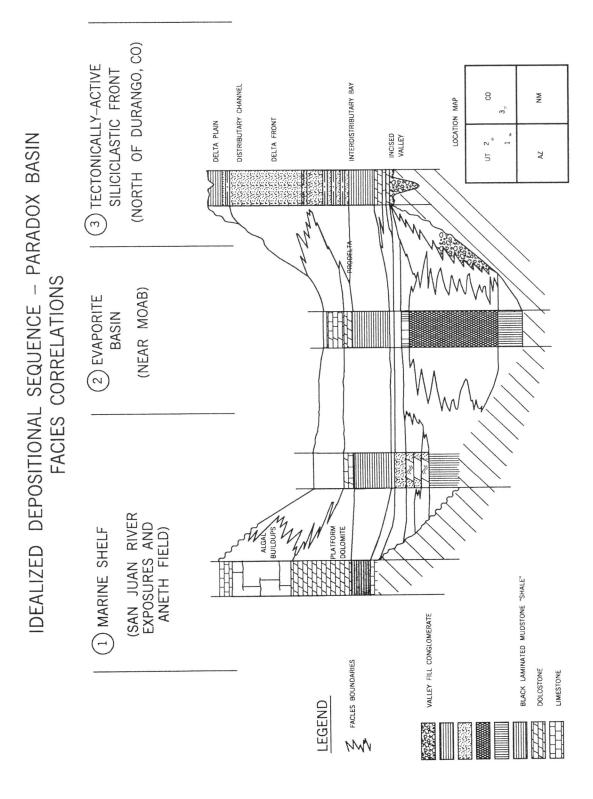

Figure 16. Idealized depositional sequence showing facies correlations.

FACIES DESIGNATION	ROCK TYPE	BEDDING/STRUCTURE	GRAIN TYPES	DISTRIBUTION/LOCATION	DEPOSITIONAL SETTING	EARLY DIAGENESIS
ANHYDRITE FACIES (AN)	Anhydrite; less common dolomudstone and cryptalgalaminite	Anhydrite interlayered with dolomudstone; vertical growth habit of anhydrite crystals	-----	Basin: topographic lows; lowstand systems tracts (up to 10 m thick)	Shallow hypersaline salinas; short-term sabkha conditions (0-10 m)	Dolomitization; evaporite pore filling; silicification
CRYPTALGALAMINITE/ STROMATOLITE FACIES (CL)	Dolomudstone to dolopackstone	Stacked, low-relief, wavy, ultra-thin laminae; desiccation cracks	Peloids; quartz silt	Basin: topographic lows; lowstand systems tracts (3-10 m thick); rarely preserved on platform during early transgression	Shallow salinas and intertidal settings (0-5 m)	Dolomitization; evaporite pore filling
SILTSTONE/SANDSTONE FACIES (SS, XPG)	Siltstone and less common sandstone	Low-angle cross-stratification; sharp erosional base; burrows; mudcracks	Peloids; tubiclasts; quartz silt and sand	Platform: thin (<1m) marine reworked sand sheets above sequence boundary or as parasequence caps; Basin: thin to thick (up to 7 m); lowstand and early TST	Platform: shallow marine reworking of eolian-derived siliciclastics; basin: spill-over sedimentation of eolian-derived quartz into topographic lows adjacent to carbonate platform	Calcite pore filling; evaporite pore filling in basin
DOLOMUDSTONE/ WACKESTONE FACIES (DM/W)	Dolomudstone and dolowackestone	Fenestrae, bioturbation; mm-scale wispy laminations separated by nonlaminated zones; mudcracks; fissures; anhydrite laths and nodules	Abraded skeletal debris; intraclasts; quartz silt	Platform: widespread TST deposits or thin and discontinuous cycle caps in HST (1-6 m thick); Basin: widespread TST and LST deposits (1-10 m thick); best reservoir quality in TST	Platform: peritidal, restricted lagoon to tidal flat (0-2 m); basin: intertidal and restricted subtidal (0-10 m)	Dolomitization; dissolution; evaporite pore filling
DOLOMITIZED PACKSTONE/ GRAINSTONE FACIES (DP/G)	Dolopackstone and dolograinstone	Inverse grading and subtle cross-stratification	Ooids, peloids, and oncoids	Platform margin: HST: prograded sand sheets filling available space; toe of slope: HST: carbonate sand aprons 1-3 km out away from platform; basin: LST: narrow fairway in basin parallel to strike of platform on windward margin	Platform margin: out of active environment of ooid formation, in 5-20 m water, storm- and wave-activated prograding carbonate sand sheets; toe of slope: sediment gravity flow deposits, 20-50 m water, called AOG when not pervasively dolomitized; basin: beach deposits onlapping carbonate platform, water depth <2 m	Dolomitization
MUDSTONE/WACKESTONE FACIES (M/W)	Mudstone and wackestone	Fenestrae, usually mud filled	Locally abundant skeletal debris	Platform: widespread TST deposits or thin and discontinuous cycle caps in HST; Basin: areally restricted LST deposits	Peritidal setting on downdip margin of tidal flats, subtidal lagoons, on low-energy tidal channels (<3 to 5 m)	Partial dolomitization
BLACK LAMINATED MUDSTONE FACIES (BLM)	Carbonaceous, silty dolomudstone or mudstone	Massive, faintly laminated	Rare skeletal debris	Widespread throughout platform and basin in early TST, 2 m thick along crestal platform, 15+ m thick in lowest depressions surrounding MCU	Pelagic deposition of phytoplankton in shallow water as a result of water chemistry changes associated with regional sea level rise and marine flooding of laterally extensive salt pans; water depth ranged from 2-50+ m	-----
OOLITIC GRAINSTONE FACIES (MOG, SOG, OG)	Grainstone and packstone	Well-sorted; current stratified; bioturbated; caliche below major exposure surfaces	Ooids, peloids, abraded skeletal debris	Localized areas on Platform during HST in upper Desert Creek	Shallow agitated current swept water; beach ridges or as shallow sand sheets and lobes (0 to 3 m)	Marine and meteoric cement, dissolution
NONSKEL/SKELETAL PCKST GRNST FACIES (PP)	Grainstone and less packstone	Moderate- to well-sorted, burrows and low-angle cross-lamination	Peloids, superficial ooids, aggregated grains, skeletal debris	Platform: cycle caps in HST and TST	Wave and current swept; restriction near cycle caps; shallow subtidal to lower intertidal (0-5 m)	Marine and meteoric cement; dissolution
SKELETAL PACKSTONE FACIES (SP)	Packstone with lesser wackestone	Burrows	Echinoderms, brachs, molluscs, fusulinids, bryozoans, encrusting forams, phylloid algae, peloids	Platform: laterally discontinuous TST and HST deposits; cap subtidal cycles along platform margin	Wave and current swept shallow subtidal (1-10 m)	Cementation
SKELETAL WACKESTONE FACIES (SW)	Wackestone with lesser packstone and mudstone	Burrows	Fusulinids, crinoids, brachiopods, ostracodes, bryozoans	Platform and basin in LST, TST, and HST	Well oxygenated; normal-marine; subtidal environment below high energy wave base (5-30+ m)	-----
PHYLLOID ALGAL BAFFLESTONE FACIES (AB)	Packstone to Boundstone (Bafflestone)	Massive, primary growth fabric	Phylloid algae, other skeletal debris, peloids	Initial depositional cycle above MFS of 3rd-order rise	Shallow subtidal, well-circulated, normal marine water (5-15 m)	Fibrous, botryoidal marine cement
PHYLLOID ALGAL PACKSTONE/ GRAINSTONE FACIES (AP/G)	Grainstone and packstone	Synsedimentary compaction	Medium to coarse fragmented and abraded phylloid algal fronds	Dominantly in HST on platform during 3rd-order rise, thin depositional units	Shallow subtidal, well-circulated, normal marine water (2-5+ m); high-energy	Dissolution of algal fronds, minor calcite cementation
PHYLLOID ALGAL WACKESTONE FACIES (AW)	Wackestone with locally abundant packstone and mudstone drapes	Synsedimentary compaction	Skeletal debris, peloids	Platform: thin discontinuous cycle caps in HST; less common in TST; basin: areally discrete "failed" buildups in HST	Shallow subtidal to intertidal, normal marine water (0-2+ m on platform; 10+ m in basin)	Dissolution; partial dolomitization of platform
ALLOCHTHONOUS OOLITIC GRAINSTONE FACIES (AOG)	Grainstone	Fine to medium-grained ooids, cross-stratified, well-preserved cortices	Ooids, peloids less common	Basinal HST deposit restricted to leeward (southern and southeastern) areas adjacent to platform; 10-13 m thick and pinch out 1-5 km away from platform	Allochthonous deposits formed by sediment gravity flow processes in 20-50 m water depth	-----
ALLOCHTHONOUS PELOIDAL/ SKELETAL PACKSTONE FACIES (APP)	Packstone and grainstone	Fine-grained, cross-stratified	Peloids, skeletal debris, less commonly superficial ooids	Basinal HST deposit as widespread sand sheet	Sediment gravity flow deposit, but hemipelagic sedimentation not ruled out (20-50 m)	-----

Table 1. Salient features of facies recognized in this study.

vicinity of the Aneth field. Present and subsequent discussions of estimates of water depth are based on 1) the spatial position of facies within genetically-related shallowing-upward cycles which are constrained by platform to basin correlation of chronostratigraphic surfaces, 2) antecedent depositional topography, 3) mapped distribution of diagenetic features (i.e., caliche and silcrete horizons, meteoric solution fabrics, vadose cements, etc.), and 4) modern analogues.

CRYPTALGALAMINITE/STROMATOLITE LITHOFACIES

The cryptalgalaminite/stromatolite lithofacies is thin- to thick-bedded, ranges from 10 to 30 feet (3-10 m) in thickness, and is dominantly located in the basin. The laminite facies is interlayered with siltstone and less commonly anhydrite (Plate 1-C). Millimeter thick, alternating black and tan, smooth flat and wavy crinkled laminations are characteristic of this facies (Plate 1-D). Layered and enterolithic anhydrite, stylolitic shaley seams, and thin rip-up intraclastic lenses are, in places, interbedded with algal laminae. Desiccation prism cracks (Plate 2-A) and burrows disrupt the lamination. Morphologies of algal lamination are flat, domal, broad arcuate, and digitate.

The dominance of ultra-thin, wavy, and parallel laminations, the presence of desiccation cracks, and the association with anhydrite and siltstone indicate that these sediments were formed in shallow salinas and in intertidal settings where algal mats flourished but periodically dried out. Domal and digitate stromatolites were probably deposited in slightly deeper water relative to cryptalgalaminites. In general, water depth probably ranged from 0 to 33 feet (0-10 m). Dolomitization and replacement by anhydrite are interpreted to have occurred syndepositionally or during shallow burial (for analogy, see Patterson and Kinsman, 1982). This facies is probably more widespread than observed, but is effectively masked by poor preservation of cyanobacteria and early diagenesis.

SILTSTONE/SANDSTONE LITHOFACIES

On the Aneth Platform, the siltstone/sandstone lithofacies consists of tan-gray, green-gray, and dark gray calcareous coarse-grained siltstone and very fine-grained sandstone (Plate 2-B). Calcareous debris is admixed and contains abraded peloids, skeletal debris, and lithoclasts (Plate 2-C). Limestone lithoclasts are composed of calichified crusts and well-cemented peloidal and oolitic grainstone with spar-filled moldic porosity. The basal contact is sharp and erosional with the underlying carbonate facies. In general, throughout the Aneth Platform siltstone/sandstone beds are laterally discontinuous and are rarely more than 3 feet (1 m) in thickness.

The siltstone/sandstone facies is interpreted to represent shallow marine reworking of eolian-derived siliciclastic material. Eolian transport occurred during subaerial exposure of the carbonate platform during lowstands of sea level. Periodic subaerial exposure is suggested by autobrecciation of thin beds and by compacted desiccation cracks. Subsequent marine incursion modified and destroyed most structures diagnostic of eolian processes.

The sandstone/siltstone facies is also present in the basin adjacent to the northern, eastern, and southern platform areas of Aneth. Fine- to coarse-grained siltstone and very fine-grained sandstone are interbedded with and gradationally overlie the cryptalgalaminite/stromatolite lithofacies and approach 16 to 23 feet (5-7 m) in thickness. These quartz clastics are dark gray to black in color and are thinly laminated with wavy to crinkly laminations (Plate 2-D).

In the basin, the siltstone/sandstone lithofacies probably originated as spill-over sedimentation of eolian-derived quartz grains into topographic lows surrounding platform areas. Silt and sand are trapped in shallow restricted saline ponds and on tidal flats. Disrupted laminae and scour surfaces suggest that these siliciclastics were reworked and redeposited as marine units.

DOLOMUDSTONE/WACKESTONE LITHOFACIES

The dolomudstone/wackestone lithofacies is present on the carbonate platform and in the basin. On the platform, dolostone varies in color from light brown to tan, whereas, in the basin this lithofacies is dark brown or dark gray-brown. The dolomudstone/wackestone lithofacies may exceed 23 feet (7 m) in thickness on the platform, but generally ranges from 3 to 20 feet (1-6 m). Two types of platform dolomudstones occur. The first is a thin (~3-6 feet) lenticular depositional unit that is correlative on a local scale (several thousand feet or less) and caps shallowing upward depositional cycles. Mud and spar filled fenestrae occur within this dolomudstone (Plate 3-A), and there is an absence of anhydrite nodules. Cycle capping dolostones overlie algally-dominated rocks. The second type of platform dolomudstone occurs on the

Aneth Platform as widespread correlative units. Silty interlayers and laminae, intraclastic laminae and beds, and rounded anhydrite nodules (1-3 cm in diameter) and crystal laths are distinguishing features found in these dolomudstones (Plate 3-B). Sedimentary structures consist of millimeter-scale wispy laminations which are separated by non-laminated zones generally less than 5 to 10 centimeters thick (Plate 3-C). Spar-filled mudcracks and fissures are recognized along the top of non-laminated zones (Plate 3-D). Moldic and intercrystalline porosity are evident, and some voids result from leached fossil allochems (Plate 4-A). Based on the shape of vugs, original grains were probably molluscs. Post-depositional dissolution of evaporites also contributes to porosity enhancement.

Mud-filled fenestrae, mudcracks, subvertical syndepositional cracks and fissures, very fine-grained dolomite crystals, and nodular anhydrite suggest this facies was deposited in a restricted, shallow water lagoon and/or peritidal setting. Water depth probably did not exceed 6 feet (2 m).

In the basin, the dolomudstone/wackestone lithofacies ranges from 3 to 33 feet (1-10 m) in thickness and is commonly over- and underlain by the anhydrite, cryptalgalaminite/stromatolite, and siltstone/sandstone lithofacies. The dolomudstone/wackestone lithofacies is not well-laminated. Burrow disruption and homogenation have commonly destroyed primary lamination. Irregular mud-filled laminoid fenestrae occur. Microcrystalline carbonate mud (now dolomite) with scattered detrital carbonate grains, such as pellets, peloids, small mud intraclasts, and rare fossil fragments characterize this lithofacies. Lenses of peloid packstone and quartz-rich clastics are interlayered with the mudstone. Intercrystalline and small-scale solution porosity is observed and is locally well-developed.

These basinal dolomudstone deposits are interpreted to have formed in a shallow subtidal to quiet, middle and lower intertidal environment. The presence of bioturbation and burrow disruption of layering and lamination, fenestrae and rare fossil fragments, and by close stratigraphic association with other shallow water rocks (e.g., anhydrite, cryptalgalaminites, stromatolites, and fine-grained quartz clastics) support this interpretation.

DOLOMITIZED PACKSTONE/GRAINSTONE LITHOFACIES

The dolomitized packstone/grainstone lithofacies is light to dark brown and pervasively dolomitized. It exhibits good reservoir quality and yields hydrocarbons. Fine-grained ooids, peloids, and/or oncoids occur in a grain-supported matrix (Plate 4-B). The dolomitized packstone/grainstone lithofacies occupies three different positions along the platform to basin transect, and is divided here into three subfacies.

Along the plaform margin in the northern and western portion of the Aneth Field, peloidal and oolitic sand were deposited down dip from the active carbonate factory during late Desert Creek time. Active formation of carbonate grains occurred in very shallow agitated water along the crest of the platform. Time equivalent platform margin sedimentation took place in deeper water 16 to 60 feet (5-20 m). Unlike the active ooid- and peloid-forming environments, these "deeper" water accumulations were relatively unaffected by pervasive marine fibrous cementation that occludes much of the interparticle porosity in updip positions.

At the toe of slope or along the basin margin where the previously described facies spilled off the platform into the basin, carbonate sand aprons formed. They extend out away from the platform for several miles. These allochthonous oolitic and peloidal grainstones result from mass sediment gravity movement. The coarsest grains are deposited adjacent to the platform and become fine-grained distally. In addition, grain size tends to coarsen upward within the carbonate sand aprons. On the platform crest within the environment of formation of these shallow-water carbonate allochems, a similar trend in grain size is observed; fine-grained peloids coarsen up and grade into fine-grained ooids which coarsen upward into medium-grained ooid grainstone. Interparticle porosity is volumetrically important in the sand apron deposits. Coarser grainstones yield larger pores and higher permeability. The basin margin dolomitized packstone/grainstone lithofacies was deposited in 60 to 120 feet (20-40 m) of water. It should be noted that where these basin margin grainstones are not pervasively dolomitized, they are referred to as the allochthonous peloidal/skeletal packstone lithofacies or allochthonous oolitic grainstone lithofacies. These facies are discussed later.

The third subfacies of the dolomitized packstone/grainstone lithofacies is unlike the first two subfacies in that it occurs in the basin as *in-situ* or autochthonous deposits. These deposits are composed of ooids, peloids, and oncoids up to several mm in grain diameter. Interparticle porosity is observed as the dominant pore type of this subfacies. These deposits have only been recognized along the northern or windward side of the Aneth Platform as a narrow fairway parallel to the strike of the platform. Deposition probably occurred on or near a beach environment in very shallow restricted water, probably less than 6 feet (2 m) in depth. The rock is pervasively dolomitized.

MUDSTONE/WACKESTONE LITHOFACIES (M/W)

The mudstone/wackestone facies is light tan-brown to gray, partially dolomitized with locally abundant skeletal debris and mud-filled fenestrae. It is found on the platform and in the basin and is usually less than 10 feet (3 m) thick. Distribution of the mudstone/wackestone facies is similar to that of the dolomudstone/wackestone facies. The mudstone/wackestone facies formed in a peritidal setting along the down dip margin of tidal flats, in shallow subtidal lagoons, or in low-energy tidal channels. Water depth is postulated to be less than 20 feet (6 m).

BLACK LAMINATED MUDSTONE LITHOFACIES

The black laminated mudstone facies has been described as black shale (Hite and Buckner, 1981; Choquette and Traut, 1963; Elias, 1963; Peterson, 1966). Although this facies is argillaceous (clay-bearing) and contains silt, it is predominantly composed of carbonate minerals and is described here as a sapropelic (organic-rich) or carbonaceous dolostone (Roylance, 1984). In the literature these dolostones (e.g., "A", "C", Chimney Rock, Gothic, Hovenweep, and Paradox) are informally referred to as "shales". A literature review suggests that the Hovenweep and Paradox "shales" do not extend across some topographic highs. The "A", "C", Chimney Rock, and Gothic occur throughout the basin. In outcrop, the black laminated mudstone facies exhibits fissility, where weathered. Along Honaker Trail, a pit has been dug in the Chimney Rock and non-weathered mudstone is massive, not fissile. The Chimney Rock has a total organic carbon (TOC) content that exceeds 15 % in some areas (Hite, 1970).

Sapropelic dolostone intervals (i.e., Chimney Rock and Gothic) are from 3 to 75 feet (1-23 m) thick and extend throughout the entire Paradox basin (Roylance, 1984). In and around McElmo Creek, the black laminated mudstone facies varies in thickness from 7 feet (2 m) along the crest of the platform to 50 feet (15 m) in bathymetric lows in the basin. These dolostones contain pyrite and organic matter (Plate 4-C) and grade up into a clean, light brown tidal flat dolomudstone of the dolomudstone/wackestone facies (Plate 4-D). Based on stratigraphic reconstructions using the chronostratigrapic surfaces in and around the Aneth Field, water depths ranged from near wave base to 175 or more feet of water. Restriction of fauna, lack of burrows, good preservation of flat lamination, and presence of pyrite suggest that prevailing chemical conditions were hostile to animal life and probably disaerobic if not periodically anoxic.

The black laminated mudstone facies is interpreted to have been deposited during regional relative rises in sea level. This facies generally follows a period of basin restriction and evaporite deposition. Initial invasion of marine water into the shallow hypersaline sea of the central Paradox Basin, caused dissolution of halite and other polysalts, and an increase in chloride concentration. Mesosaline water prevented stenotopic or even most eurytopic organisms from existing. Cyanobacteria were perhaps the only organisms capable of surviving such harsh conditions. Low diversity and high abundance of bacteria resulted in bacterial blooms which sedimented the sea floor throughout the Paradox Basin. In time, bacterial/organic ooze draped the central Paradox salt basin and prevented further wholesale dissolution of evaporites. Normal marine circulation patterns eventually diluted the chloride concentrations and returned the basin to normal marine conditions.

High organic production within mesosaline water (salinity range of 4 to 12 percent, 3.5 percent representing normal-marine) is due to the proliferation of phytoplankton (Kirkland and Evans, 1981). Some phytoplankton survive high salinities, pH variations, low carbon dioxide concentrations, and other conditions hostile to larger organisms. They flourish in a constant and concentrated supply of nutrients resulting from the evaporation and replenishment of seawater. They also flourish where grazers and parasites are absent.

The high salinities promote anoxic conditions and organic preservation. The solubility of oxygen decreases very rapidly with increasing salinity (Copeland, 1967; Hite, 1970; Kinsman and others, 1973; and Kirkland and Evans, 1981). The remaining oxygen is rapidly consumed by oxidation of abundant organic matter and hydrogen sulfide (Kirkland and Evans, 1981). The excess organic matter accumulates in the sediment. Two factors kept this anoxic layer undisturbed by wave or current action. The limited fetch of the small Paradox basin, and the fact that saline water is denser than normal-marine water and is more difficult to move by wind, resulting in limited wave action.

Mesosalinity and rapid sedimentation of organics resulted in anoxic or dyserobic bottom conditions and allowed the preservation of organic matter. This organic ooze later became the source of hydrocarbons and the reservoir seal for the Greater Aneth Field and other oil fields producing from the Middle Pennsylvanian in the southwestern shelf area of the Paradox Basin.

OOID GRAINSTONE LITHOFACIES

The ooid grainstone facies is observed only on the platform and is restricted primarily to the uppermost Desert Creek. This facies ranges in thickness from 0 to 33 feet (0-10 m) and is generally divided into a lower stabilized sand flat subfacies and an upper active or mobile sand belt subfacies. The lower 1/3 to 1/2 of the ooid grainstone facies is moderately sorted, bioturbated, and grain-supported with disseminated carbonate mud – now micrite (Plate 5-A). A sharp contact separates the lower portion from the upper well-sorted, current stratified, mud-free ooid grainstone (Plate 5-B). Throughout this facies, ooids increase in size from base to top; ooids rarely exceed 1 mm in diameter. Calcretes, silcretes, vadose silt, and vadose cements occur toward the top of this facies and indicate conditions of subaerial exposure. Marine fibrous cements are ubiquitous and occlude much of the interparticle pore network. Active marine cementation is most conspicuous in the mobile sand belt subfacies. Post-depositional meteoric leaching of ooids during periods of platform exposure has caused oomoldic porosity; porosity exceeds 30% in some intervals, but permeability is generally less than 10 md.

Ooid buildups within the Desert Creek show broadly similar subenvironments to the modern. Shoal development initiated on a raised topographic surface (crestal platform areas) above fairweather wave base. Here, ooids formed and accumulated in an area of wave and current agitation (based on modern analogs, water depth was less than 10 feet). Midway through development of the ooid grainstone facies, ooid formation ceased for a time. Bioturbation and minor mud infiltration occurred. Subsequently, the ooid forming carbonate factory resumed. Within the upper portion of the facies, tides and waves produced cross-stratified bedforms which migrated down dip accumulating over much of the preexisting carbonate platform. In some areas ooids were transported off the platform by mass-sediment-gravity movement.

The active ooid forming environment was located on crestal positions of the Aneth platform. Active ooid formation took place in shallow agitated current-swept water, 3 to 10 feet (1-3m) in depth. Most ooids underwent minor transport and were deposited in isolated subaerially exposed beach ridges or in shallow subtidal sand sheets and lobes across the platform.

NON-SKELETAL/SKELETAL PACKSTONE/ GRAINSTONE LITHOFACIES

The non-skeletal/skeletal packstone/grainstone facies forms on the platform as shoal water cycle caps that range from 5 to 20 feet (2-7 m) in thickness. This facies consists of moderately well-sorted, fine- to coarse-grained carbonate sand grainstone and lesser packstone. Non-skeletal allochems include peloids, superficial ooids, aggregate grains, and reworked rounded intraclasts. Skeletal material is present and includes crinoids, forams, brachiopods, molluscs, bryozoans, and phylloid algae. Micritized skeletal grains are common. Sedimentary structures include burrows and ripple cross-lamination. The basal contact of this facies is gradational above deeper water facies (e.g., skeletal wackestone and packstone facies), and the upper contact is sharp and displays evidence of subaerial exposure (i.e., caliches, silcretes, rhizoliths, alveolar textures, silt filled subvertical fissures; Plate 5-C and 5-D). Solution-enhanced interparticle porosity is well-developed; skeletal debris and peloids have been leached forming moldic pores that are partly to completely filled with coarse blocky calcite cement.

Grain-support, abraded and micritized skeletal debris, abundant encrusting forams, and the light color of the non-skeletal/skeletal packstone/grainstone facies indicates shallow subtidal to lower intertidal water depths 0-15 feet (0-5 m). This interpretation is based on analogues from the Holocene (Harris, 1979). Hard peloid sand and mechanically-generated cross-stratification suggest that this facies underwent deposition in high-energy shoal water conditions. Equant crystal morphology of pore filling cement within skeletal molds strongly suggests that cementation took place within the meteoric phreatic diagenetic environment (Peterson and Hite, 1969). Extensive fresh water leaching is linked to subaerial exposure and the development of a meteoric water lense (Peterson and Hite, 1969; Herrod and others, 1985).

SKELETAL PACKSTONE LITHOFACIES

The skeletal packstone facies is recognized on the platform and in the basin as 1 to 10 feet (<1 to 3 m) thick depositional units. This facies consists of very fine- to medium-grained skeletal packstone with lesser wackestone and grainstone. Normal-marine skeletal allochems characterize the facies and include crinoids, echinoids, trilobites, brachiopods, molluscs, fusulinids, bryozoans, encrusting forams (opthalmidids), and phylloid algae (Plate 6-A and 6-B). Non-skeletal grain types occur mainly as peloids. The skeletal packstone facies commonly overlies skeletal wackestone and is in turn overlain by the non-skeletal/skeletal packstone/grainstone facies. The skeletal packstone facies may form the uppermost depositional unit of shallowing upward cycles.

Shallow subtidal to near shoal-water conditions are postulated for this facies. Water depths ranged from periodically emergent to perhaps 30 feet (10 m). Modern encrusting opthalmidid forams studied in Florida Bay are most abundant in water that range in depth from 0 to 15 feet (0-5 m) (Elias, 1963). Grain-supported rock texture, diverse faunal assemblage, lack of abundant detrital clastics lend support that this facies formed in open marine, subtidal environments consistent with the water depths indicated above.

SKELETAL WACKESTONE LITHOFACIES

The skeletal wackestone facies is divisible into two grossly similar, but genetically different non-reservoir subfacies. The dark skeletal wackestone subfacies is dark gray to black, <1 to 3 feet (<1 m) thick and is stratigraphically juxtaposed to the black laminated mudstone facies. This skeletal wackestone may appear shaley (Plate 6-C). Petrographic investigation indicates that the dark color is a result of disseminated organic matter and sulfides, mostly pyrite. Articulated, fragmented, and crushed echinoderm and brachiopods are the dominate allochems. Skeletal material is disseminated in a mud matrix that is commonly dolomitized (Plate 6-C).

The dark skeletal wackestone subfacies is interpreted to have formed in a subtidal setting related to water chemistry changes associated with the black laminated mudstone facies. The occurrence of *in situ* stenotopic fauna (e.g., echinoderm and brachiopods) suggests that marine conditions were normal or nearly so. However, the occurrence of disseminated organics and pyrite suggests that sediment just below the sediment-water interface was reducing and probably dysaerobic or anoxic. The dark skeletal wackestone subfacies formed during the regional sea level rises that resulted in deposition of the black laminated mudstone facies. On the Aneth Platform tidal flats of the dolomudstone/wackestone facies gradationally overlie the dark skeletal wackestone subfacies. Water depth at deposition was shallow.

The second subfacies is a medium gray skeletal wackestone that occurs both on the platform and in the basin as basal depositional units within shallowing upward cycles. This subfacies ranges from 1 to 10 feet (<1 to 3 m) in thickness, and is medium-gray with a normal marine biota, consisting of forams, crinoids, brachiopods, ostracodes, branching and fenestrate bryozoans (Plate 6-D). This subfacies was deposited in a normal-marine, well-oxygenated, subtidal environment below normal wave base. This interpretation is supported by the abundance of normal marine biota, mud-supported rock fabric, light color, paucity of quartz silt, and the general lack of sulfides and organic matter. Water depths are postulated to have been approximately 15 feet (5 m) on the platform to perhaps as much as >100 feet (30+ m) in the basin adjacent to the carbonate platform.

PHYLLOID ALGAL LITHOFACIES

This lithofacies is divided into three subfacies that are recognized as grossly similar in appearance, yet genetically they occupy different environmental niches, which influence post-depositional alteration and reservoir quality. The phylloid algal bafflestone subfacies has a packstone to boundstone (bafflestone; after Embry and Klovan, 1971) texture (Plate 7-A) and is restricted in occurrence to the core of the Aneth Platform. Where present, it accounts for the lowermost 10 to 15 feet (3-5 m) of the carbonate platform. This subfacies consists of phylloid algal plates that are separated by syndepositional, botryoidal-fibrous cement (Plate 7-B). The marine cement contributed to an early, rigid framework. Phylloid algal plates provided the solid substrate required for marine cement nucleation, while shelter cavities provided space for crystal growth. Mud and fine skeletal debris are restricted to sediment traps or pockets. Primary porosity is volumetrically important within this facies; interparticle and shelter porosity is formed by irregular packing of algal plates.

The phylloid algal packstone/grainstone subfacies is characterized by fine to coarse fragmented and abraded phylloid algal fronds (Plate 7-C). This facies is recognized on the platform as thin depositional units usually less than 10 feet (3 m) thick. Good interparticle porosity is developed (Plate 7-C), and secondary algal moldic porosity and solution enhanced interparticle porosity is common. Choquette and Traut (1963) indicated that much of the secondary dissolution and pore enhancement occurred early in the diagenetic history of these rocks.

The phylloid algal wackestone subfacies is found on the platform as thin wackestone and packstone layers (packstone may be locally abundant) with intercalated carbonate mud drapes (Plate 7-D). Observed thicknesses vary from 3 to 15 feet (1 to 5 m). Within this subfacies solution voids cross-cut depositional fabric. Porosity is highly variable and depends on texture and degree of dolomitization. Algal moldic porosity is volumetrically more abundant than interparticle porosity. Porosity formation is related to subaerial exposure of the tops of cycles containing algal bioherms and concomitant dissolution of less-stable carbonate mineralogies by meteoric water during early diagenesis (Peterson and Hite, 1963; Hite, 1970; Wilson, 1975; Hite and Buckner, 1981; Choquette, 1983; Herrod and Gardner, 1988; Dawson, 1988).

A normal marine fauna, dominance of photosynthetic algae, and the light color of these three subfacies suggests that deposition occurred in shallow subtidal, well-circulated, normal marine water. Stratigraphic cross-sections within a chronostratigraphic framework suggest paleo-water depths extended from strandline (shoreface) to 50 feet (15 m) of water. These facies formed within a broad partially overlapping depositional spectrum. In an ideal depositional cycle from deeper to shallower water, algal boundstone is overlain by algal packstone/grainstone, which is overlain by partially dolomitized algal wackestone. It should be noted that along the platform margin and in the basin, where platform development lagged or ultimately failed, the algal wackestone subfacies probably formed in deeper water ~30 feet (~10 m). Dolomitization is not generally recognized in these down dip locations. Porosity is poorly developed in the deeper water algal wackestone accumulation.

ALLOCHTHONOUS OOID GRAINSTONE LITHOFACIES

The allochthonous ooid grainstone facies is a basinal deposit primarily restricted in occurrence to the leeward (southern and southeastern) margin of the Aneth platform. These carbonate sand apron accumulations generally do not exceed 30 to 40 feet (10-13 m) in thickness and pinch out 1-4 miles from the platform. This facies is light to dark gray-brown and is composed of cross-stratified fine- to medium-grained ooids with well preserved ooid cortices (Plate 8-A and 8-B). Ooid grain size mimics grain size observed in the ooid-forming environment on the platform. Unlike the penecontemporaneous ooid grainstone deposited on the platform, these ooids were not exposed to long-term subaerial conditions. Porosity is well-developed in the allochthonous ooid grainstone facies, especially in positions adjacent to the platform where coarser grains result in better connection of interparticle pores and better permeability. Dolomitization of ooids in the allochthonous ooid grainstone facies is minimal except near the contact with overlying pervasively dolomitized rocks. Preserved ooid cortices and calcitic mineralogy are used to differentiate the allochthonous ooid grainstone facies from the dolomitized packstone/grainstone facies. The allochthonous ooid grainstone facies formed in 60 to 150 feet (20-45 m) of water in the vicinity of the Aneth platform as a result of gravity flow processes. Postulated water depths are based on platform relief during upper Desert Creek time. The spatial position of grainstone facies is constrained by platform to basin correlation of chronostratigraphic surfaces. These deposits were unaffected by marine fibrous cementation.

ALLOCHTHONOUS PELOIDAL/SKELETAL PACKSTONE LITHOFACIES

The allochthonous peloidal/skeletal facies is similar to the allochthonous ooid grainstone facies in that both were deposited in late Desert Creek time under similar depositional regimes. Differences between the two lithofacies exist in grain type and size, distribution, and geometry. The allochthonous peloidal/skeletal packstone facies is composed of very fine-grained peloids, skeletal debris, and less commonly superficial ooids in a grain-supported matrix (Plate 8-C and 8-D). This lithofacies rarely exceeds 20 feet (6 m) in thickness and is the basin equivalent to the non-skeletal/skeletal packstone/grainstone facies and skeletal packstone facies. In the basin, the allochthonous peloidal/skeletal packstone facies is overlain by the allochthonous ooid grainstone facies. Likewise, on the platform the non-skeletal/skeletal packstone/grainstone and skeletal packstone/grainstone facies are overlain by ooid grainstone. The allochthonous packstone facies is a widespread carbonate unit with sand sheet geometry. Unlike the allochthonous ooid facies, this facies is not restricted to the leeward side of the carbonate platform. Mass-sediment gravity movement is postulated. However, due to minute grain size and widespread distribution, hemipelagic sedimentation is not ruled out. Water depths range from 60 to 150 feet (20 to 50 m).

MIDDLE PENNSYLVANIAN (DESMOINESIAN) STRATIGRAPHIC FRAMEWORK

In this section we apply sequence stratigraphic concepts to the Middle Pennsylvanian strata of the Paradox basin region. Integration of outcrop, subsurface core, well log, biostratigraphy, and seismic data into a sequence stratigraphic framework has provided insight into the exploration and reservoir-scale stratigraphy of these important hydrocarbon bearing units. Stratigraphic concepts and terminology and the Desmoinesian sequence stratigraphic framework are presented below.

SEQUENCE STRATIGRAPHIC NOMENCLATURE AND THE STRATIGRAPHIC HIERARCHY

Sequence stratigraphy defines stratal units that develop in response to changes in shelfal accommodation. These packages of rock are bounded by specific stratal discontinuity surfaces that can be identified in outcrop and on well logs, and where sediments are thick enough, on seismic data. These surfaces are basinwide in extent and are associated with significant changes in facies architecture. Sequence boundaries can be dated biostratigraphically, and sequences can be used as chronostratigraphic units if the bounding unconformities are traced to the minimal hiatus at their conformable position in the basin. A sequence represents all the rocks deposited in the interval of time between the age of the two unconformities. The rocks above the sequence boundary are always younger than the rocks below (Sloss, et al., 1949; Sloss, 1963; Wheeler, 1958; Vail et al., 1991). Sequence boundaries are always present in the rock record, although, in places, they may be subdued or one boundary may be a composite of several sequence boundaries.

In general, sequence boundaries are regional onlap surfaces. In basin areas they are characterized by onlap of allochthonous deposits (e.g., debris flows, slump deposits, gravity flow sand deposits), prograding deltas, carbonate platform deposits, or evaporites (Vail et al., 1991). In shallow water or non-marine settings they are characterized by onlap of strata deposited in peritidal environments of carbonate platforms, strata deposited as sabkha or shallow subaqueous evaporites (Sarg, 1988), or strata deposited in deltaic, coastal or fluvial environments. Subaerial and submarine erosional truncation are commonly present below a sequence boundary. Subaerial truncation is accompanied by varying degrees of microkarst, paleosol, and/or caliche development. Abrupt facies truncation or dislocation is also commonly present at sequence boundaries. Toplap, indicative of sediment bypassing, is a common pattern found below sequence boundaries in areas of rapid progradation.

Each depositional sequence is composed of systems tracts (Posamentier et al., 1988; Jervey, 1988). A systems tract is a set of linked contemporaneous depositional systems (Brown, 1969). Each systems tract is bounded by a physical surface that is, in part, a discontinuity marking the boundary of a similar set of accommodation patterns such as sigmoidal to oblique, oblique to aggradational, or backstepping. Depositional systems within each systems tract are linked by changes in sedimentary facies (Vail et al., 1991).

Sequences or composite sequences (i.e., groups of higher-order sequences, Mitchum and Van Wagoner, 1991) generally develop on a time scale of 0.5-5 My (3rd-order) (Figure 20). Third-order cycles tend to group into longer-term second-order cycles bounded by major relative falls in sea level. In general, a set of five to seven third-order cycles forms a second-order cycle with a duration averaging 9-10 My. The boundaries of second-order cycles are characterized by especially large falls in sea level (greater than 150 feet; 50m). The stratigraphic signature of a second-order cycle is a supersequence (Haq et al., 1987, Vail et al., 1991). The Desmoinesian of the Paradox basin and the Mid-Continent region, represents a major component of a second-order supersequence (Figure 21, oversized, see note at back).

The Desmoinesian is made up of 3rd-order composite sequences composed of lowstand, transgressive, and highstand systems tracts or sequence sets that are composed of higher-order sequences. All the Desmoinesian sequences and composite sequences are Type 1 sequences. The Paradox depositional sequence model is summarized in Figure 22. At the base of the lowstand systems tract (LST) is a Type 1 sequence boundary, characterized by subaerial exposure of the shelf, incision of the shelf if there is a fluvial system, and/or erosion on the slope. In the central and southern portion of the Paradox basin, basinally restricted wedges composed of evaporites onlap Type 1 sequence boundaries and overlie highstand systems tracts comprised of foreslope and basinal carbonate rocks. The physical boundary between the lowstand and transgressive systems tracts is defined by the change from forestepping to backstepping and is called the transgressive surface (TS). In the Paradox strata, this occurs where basin restricted siltstone changes upward to skeletal lime mudstone/wackestone. This surface merges with the basal unconformity landward of the point where the lowstand

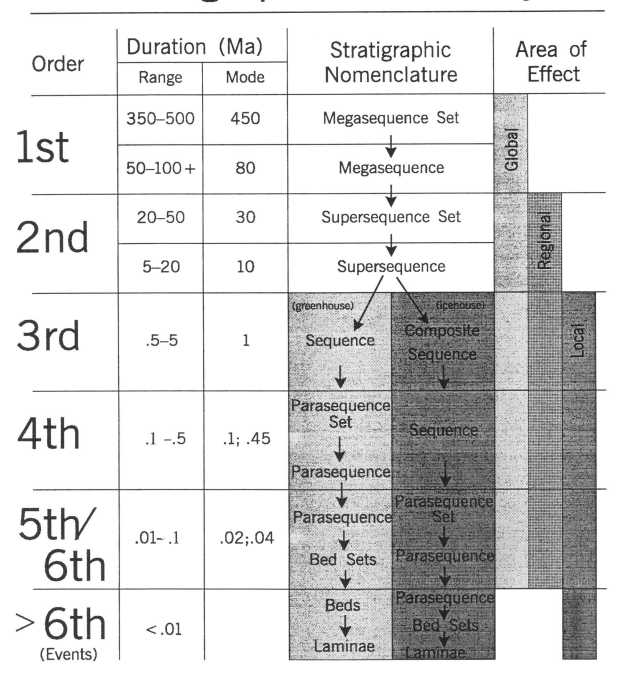

Figure 20. Orders of stratigraphic hierarchy with associated duration, nomenclature, and area of effect.

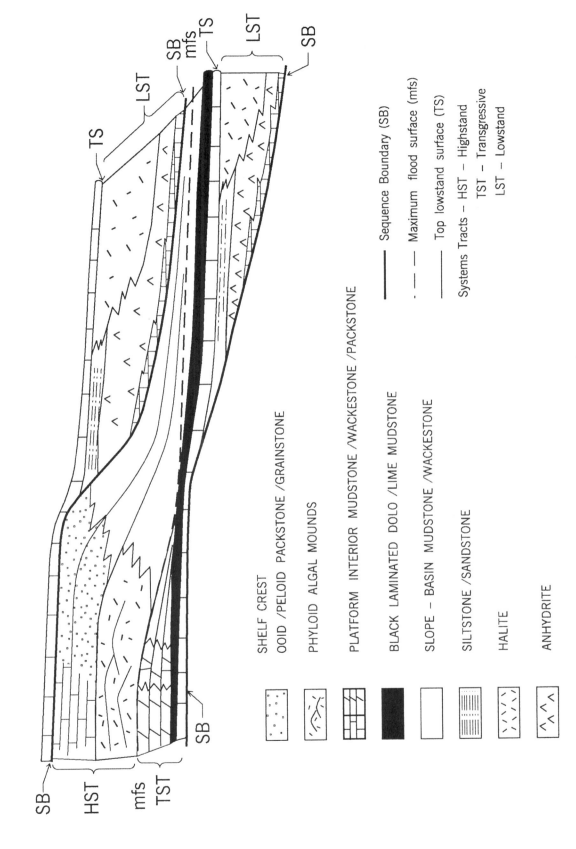

Figure 22. Sequence stratigraphic model for Desmoinesian strata, Paradox basin.

systems tract pinches out. The physical boundary between the transgressive (TST) and highstand (HST) systems tracts is the maximum flooding surface (mfs). In basinward positions, it is contained within a marine condensed section. In starved areas the maximum flooding surface is associated with a hardground or deep marine dissolution surface. In shelfal or shallow platform settings of the Paradox Desmoinesian, it is represented by the top of a silty dolomudstone/wackestone unit that overlies a transgressive black laminated mudstone. The maximum flood surface occurs at the base of open-marine, algal-rich carbonate platform facies.

The stratigraphic units deposited within fourth-order cycles are depositional sequences and fifth-order cycles or parasequences (Figure 20). Parasequences are the building blocks of fourth-order systems tracts (Van Wagoner et al., 1988, 1990). Parasequences influence reservoir and interwell heterogeneities in oil and gas fields (e.g., the Aneth field). Higher frequency sequences and parasequences develop on a time scale of 0.1-0.5 My (4th-order), and 0.01-0.1 My (5th/6th-order), respectively. A parasequence is a genetically related, relatively conformable succession of beds and bedsets bounded above and below by surfaces across which there is a distinct facies break formed by water depth increase (modified from Van Wagoner et al., 1988). In this study, the upper composite sequence (third-order) of Desmoinesian age is divided into four high-frequency sequences (fourth-order) (Figures 21 and 23, oversized, see note in back). The high-frequency sequences are subdivided into a larger number of parasequences.

Parasequences and sequences may be episodic or periodic. Episodic parasequences are caused, for example, by tidal flat migration or delta lobe shifts. They are limited in distribution and are of very short duration (usually less than 10,000 years). Periodic parasequences are characterized by regional continuity and by systematic changes in thickness between high-frequency cycles within a stratigraphic section. If the environment is shallow marine, they often correlate for hundreds of miles across different depositional settings (Vail et al., 1991).

Deepening upward or shallowing upward stacking successions of parasequences in shelf or platform settings, when combined with stratal geometry's of onlap and downlap help define the different systems tracts of a sequence. Parasequences or high-frequency sequences will aggrade or backstep in the transgressive systems tract. Parasequences tend to aggrade during the early highstand, and as accommodation space becomes less during the late highstand, they will prograde. High-frequency sequences commonly develop during the late highstand, and can make identifying the third-order sequence boundary difficult as there will be a number of candidates in shelfal areas. Regional correlation into the basin to the base of the lowstand systems tract generally resolves this problem.

Periodic parasequences are believed to be caused by climatic fluctuations associated with Milankovitch scale orbital cycles (less than 500 Ky). The Milankovitch orbital cycles have dominant periods of approximately 20, 41, and 100 Ky. These orbital cycles influence the amount of solar energy received on the earth's surface and thus affect climate. It is believed that these climatic variations induce changes in continental ice volumes, which cause eustatic changes and consequently small relative changes of sea level. When high-frequency cycles develop as sequences they nest into the systems tracts of 3rd-order composite sequences. This is particularly characteristic of icehouse times, such as the Pennsylvanian-Early Permian or the Neogene (Figure 20). During ice-house times high-amplitude glacio-eustatic changes in sea level occur at high frequencies and dominate the stratigraphic signature of third-order composite sequences.

The following section summarizes the large-scale 2nd- and 3rd-order stratigraphic framework for Desmoinesian strata. The hierarchial stacking of 3rd-order composite sequences into a Desmoinesian supersequence controls the reservoir architecture of the Aneth field area. A following discussion will present the high-frequency 4th- and 5th-order sequence architecture, and its application to subsurface reservoir architecture and delineation.

DESMOINESIAN HIERARCHY, PARADOX BASIN

The sequence stratigraphic framework for the southern and southwestern portion of the Paradox Basin is based on the recognition of regionally significant sequence boundaries that contain evidence for subaerial exposure. These sequence boundaries can be identified and correlated on well logs and seismic sections from the outcrop area of the Goosenecks of the San Juan River through the subsurface to the Aneth field area and into the basin to the north. In addition to subaerial exposure, the following criteria were used to recognize major sequence boundaries in outcrop and the subsurface: 1) sequence boundaries are onlapped by thick evaporite wedges (i.e., Alkali Gulch, Barker Creek, Akah/Desert Creek, and Ismay); the Alkali Gulch and Akah evaporite wedges are thick enough to be seen to onlap on seismic data, 2) regionally extensive black laminated mudstone occurs just above sequence boundaries in shelf areas, and 3) significant aggradational growth of carbonate platforms occurs in the early parts of sea level highstands. Sequence bounding surfaces are correlated over approximately 2000 square kilometers in southeastern Utah and are inferred to correlate over the entire Paradox basin (Figure 24).

The major sequence bounding surfaces divide the Desmoinesian into five major composite sequences. Desmoinesian age estimates range from 4 My (Odin and Gale, 1982) to 10 My (Van Eysinga, 1975). Harland and others (1989) estimate the Desmoinesian time span at approximately 8 My. This provides an age range for each of these five composite sequences of between 800,000 Ky and 2.0 My with a mean of 1.4 My, indicating that they are most probably 3rd-order composite sequences. The five composite sequences are composed of systems tracts that contain higher frequency 4th-order sequences, and 5th-order parasequences (meter-scale shoaling upward cycles).

On a larger scale, the five depositional composite sequences of the Desmoinesian section occur within a 2nd-order transgressive-regressive supersequence (Figures 21 and 23). The lower two composite sequences of the supersequence are characterized by progressive backstepping toward the basin margin, and include the Pinkerton Trail and Barker Creek formations. The upper three composite sequences are characterized by aggradation (Akah and Desert Creek formations) to progradation (Ismay and Honaker Trail formations).

The Pinkerton Trail Formation represents the transgressive systems tract of the oldest composite sequence and represents the first marine transgression over the nonmarine Molas Formation. With this transgression, shallow water carbonate platform facies were deposited over the basin area (Figure 23). The lower part of the Barker Creek Formation is interpreted to be the highstand systems tract for this composite sequence. The top of the highstand and the first Desmoinesian 3rd-order sequence boundary (DS10) occurs just below the "C" shale in platform areas. This sequence boundary correlates to the base of the Alkali Gulch evaporites in the basin (base of cycle 29 of Hite and Buckner, 1981; modified by Ramussen).

The Alkali Gulch evaporite cycles are interpreted to represent a single 3rd-order lowstand wedge deposited during a long term drawdown of the basin, and are the basal systems tract for the DS10 composite sequence (Figure 21). The lower Barker Creek highstand was subaerially exposed during the entire time of deposition of this evaporite succession. The lowstand evaporite wedge is itself composed of a series of high-frequency 4th-order cycles (cycles 20-30, Hite and Buckner, 1981). These are composed of anhydrite/halite/silty dolomite (units B-D of Hite and Buckner, 1981) and black shale to black laminated mudstone (unit A of Hite and Buckner, 1981). The evaporite/silty dolomite units are interpreted to be lowstand drawdown units, and the black shale units represent times of basin freshening. The transgressive/highstand systems tracts for the DS10 composite sequence are composed of the "C" shale and upper Barker Creek. The upper sequence boundary of this composite sequence occurs at the base of the "A" shale and correlates to the base of the Barker Creek evaporites in the basin.

The Barker Creek evaporites are also a single 3rd-order lowstand evaporite wedge and contain cycles 11 through 18 of Hite and Buckner, 1981. They are interpreted to form the basal lowstand unit of the DS20 composite sequence (Figure 23). The transgressive/highstand systems tracts of this composite sequence comprise the "A" shale and the Akah Formation. Outcrop data suggests that the maximum flood for the Desmoinesian - L. Missourian supersequence occurs within the lower Akah Formation and represents the turnaround in this 2nd-order transgressive-regressive cycle from backstepping to aggradation. Carbonate platform aggradation in the Akah highstand provides the underlying paleotopographic high for deposition of the ensuing Desert Creek phylloid algal bioherms and sand shoals.

The next two composite sequences of the Desmoinesian supersequence contain significant reservoir facies and represent the upper aggradational and first progradational composite sequences of this second-order transgressive-regressive cycle. The DS30 composite sequence is composed of a basal lowstand systems tract represented by the Akah evaporite cycles (Cycles 6-9), and overlying transgressive/highstand systems tracts, represented by the Chimney Rock and the platform carbonates of the Desert Creek (Figure 21). Outcrop and core data suggest that the maximum flood surface of this composite sequence separating the transgressive and highstand systems tracts occurs at the top of a silty dolomudstone/wackestone. The maximum flood surface of this composite sequence is overlain by coalesced algal mounds forming the main buildup phase of the Desert Creek. The phylloid algal buildups occur as the initial aggradational portion of this highstand systems tract; and are overlain at the platform margin or shelf crest by ooid/peloid grainstones (Figure 22).

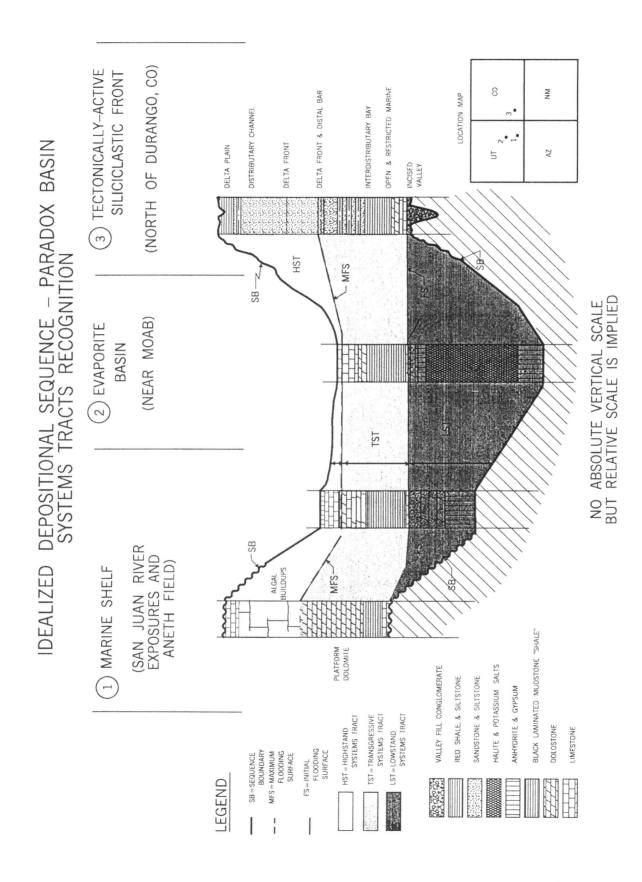

Figure 24. Idealized depositional sequence showing correlation of systems tracts.

Desmoinesian sequence 40 (i.e., DS40), the first progradational composite sequence of the Desmoinesian supersequence is interpreted to begin with deposition of evaporite cycle 3. This lowstand system is succeeded by the transgressive and highstand systems tracts of the Gothic shale, and the Ismay and Lower Honaker Trail formations. The final composite sequence of the Desmoinesian, DS50, is represented in the study area by the transgressive Ute shale and the cyclic marine limestones and clastics of the highstand upper Honaker Trail Formation.

Depositional composite sequences DS40 and 50 are dominated by cyclic 4th-order sequences each with a relatively thin evaporite lowstand unit (Figure 23). These 4th-order lowstand evaporite units are represented from oldest to youngest, by the basinal cycles 3, 2, 1, 0, 00, 000, 0000, and UHA (cycle naming convention after Hite, 1960 and Rasmussen, verbal communication 1994). The dominance of 4th-order cyclicity in the progradational component of the 2nd-order transgressive-regressive cycle indicates that the basin area was sufficiently shallow to allow restriction and evaporite deposition during 4th-order lowstands.

CORRELATION TO MIDCONTINENT

To test the hypothesis that the 3rd-order composite sequences in the Paradox basin are inter-regional in character, a tentative correlation to the Ross and Ross (1987a) coastal onlap chart and to the Midcontinent cycles of Heckel (1986) has been made using the biostratigraphic zonation presently defined for the Paradox basin and the North American Midcontinent (Figure 21). Identification of fusulinids, *Beedeina haworthi* and *B. weintzi*, in the Ismay strata, and, *Beedeina novamexicana* and *Wedekindellina* sp., in the Desert Creek strata (Baars et al., 1967) provide a biostratigraphic tie point into the North American foraminiferal zones (Ross and Ross, 1987b) (Figure 21). These tie points place composite sequences DS30 and DS40 in the upper Cherokee and lower Marmaton groups. The six 4th-order cycles within DS30 and DS40 (cycles 0, 1, 2, 3, 4, and 5) match the six widespread cycles on the Ross and Ross (1987) coastal onlap curve within these foraminifera zones. The uppermost Atokan, DS10, and DS20 composite sequences are matched to the underlying composite sequences on their cycle chart. The final composite sequence, DS 50, also shows excellent correspondence to the coastal onlap curve.

To correlate to Midcontinent cycles (Heckel, 1986), the same tie point is used and major 3rd-order transgressive "shales" of the Paradox basin are correlated to major transgressive shales of the Midcontinent. The Chimney Rock and Gothic are interpreted to correlate to the equivalent major transgressive Verdigris and Lower Fort Scott cycles (Excello shale) of the Midcontinent (Heckel, 1986). The overlying Ute shale is correlated to the major transgressive Exline cycle in the midcontinent region.

The 4th-order Hatch and Hovenweep "shales" of DS40 are interpreted to correlate to the Pawnee (Anna) and Upper Fort Scott (Little Osage) transgressive units, respectively. Two unnamed shales corresponding to the 4th-order transgressive "shales" of cycles 0 and 4 are correlated to the Altamont (Lake Neosho) and Post-Bevier of the midcontinent region. There is not a one for one match of Paradox 4th-order transgressions with all of the midcontinent transgressions. The larger number of Midcontinent transgressive shales may represent additional cycles that have not been identified in the Paradox, or they may represent higher frequency 5th-order cycles that are recognized in the midcontinent (Figure 21). Generally, there is good correspondence between the 3rd-order transgressions of this study and the major Midcontinent transgressions interpreted by Heckel (1986). Thus, it is probable, that the 3rd-order composite sequences and the 4th-order sequences identified in this study have inter-regional significance.

SEQUENCE STRATIGRAPHIC ANALYSIS: REGIONAL FRAMEWORK OF GROSS RESERVOIR SECTION

Nineteen discrete and mappable high-frequency depositional cycles (parasequences; 0.01-0.1 My) are recognized within three fourth-order (0.1-0.5 My) depositional sequences of the Desert Creek and lower Ismay section (Middle Desmoinesian) at the McElmo Creek Unit of the Aneth field, southeastern Utah. These sequences stack into parts of two third-order (0.5-5 My) composite sequences.

Desmoinesian exposures along the San Juan River and selected core, located in the Greater Aneth Field (Aneth and McElmo Creek units), were used to identify shoaling-upward depositional cycles. Intervals of significant meteoric (vadose and phreatic) diagenesis were identified and described. The following criteria are used to recognize sequence boundaries (Figure 25).

1. <u>Truncation</u>. Truncated and eroded limestone overlain by the siltstone/sandstone facies.

2. <u>Microkarst</u>. Irregular crenulated and scalloped surface in limestone exhibiting downward-directed cracks and fissures that are filled with siliciclastics of the overlying siltstone/sandstone facies.

3. <u>Retrogradational Parasequences</u>. Backstepping parasequences (upward-deepening facies) occur above sequence boundaries and are composed of the dolomudstone/wackestone facies overlain by the skeletal wackestone facies.

4. <u>Paleosols/Caliches</u>. Ooid grainstone facies and nonskeletal/skeletal packstone/grainstone facies capped by paleosols and caliches that are overlain by lenticular siliciclastics of the siltstone/sandstone facies.

5. <u>Sheet or Lenticular Siltstone/Sandstone Facies</u>. Marine reworked eolian siltstone with disseminated limestone clasts (leached ooids and caliche) overlie carbonates which display evidence of subaerial exposure.

6. <u>Pervasive Meteoric-Phreatic Diagenesis</u>. Pervasive moldic and vugular porosity and recrystallization caused by extensive meteoric-phreatic diagenesis.

7. <u>Regionally Correlative Exposure Surfaces</u>.

8. <u>Basinally Restricted Evaporites and Carbonates which Onlap Margins of Carbonate Platforms</u>.

The most diagnostic criterion for recognition of the maximum flooding surfaces (mfs) involves an abrupt change in facies and mineralogy, probably related to changing climatic and basin water conditions. Restricted-marine tidal flat dolostones are overlain by autochthonous open-marine skeletal and algal limestone facies, which are deposited in "deepest" water. It should be noted that water depths along the crest of the Aneth Platform probably did not exceed 30 to 50 feet (9-15 m) throughout deposition of the Desert Creek and lower Ismay section. The mfs's are situated just above the uppermost occurrence of the dolomudstone/wackestone facies.

Based on interpretation of cyclic stratigraphy, outcrop and subsurface stratigraphic sections are divided into regionally correlative depositional sequences. Low-frequency third-order composite sequences (e.g., DS 30 and DS 40) are comprised of higher frequency fourth-order sequences (e.g., lower Desert Creek, upper Desert Creek, lower Ismay, middle Ismay, upper Ismay, and lower Honaker Trail) (Figure 26, oversized, see note in back). The lower Desert Creek, upper Desert Creek, and lower Ismay sequences are discussed in detail below. Each fourth-order sequence reveals a complete suite of systems tracts. These nested orders of cyclicity have a profound effect on facies development and resultant reservoir performance. Third-order events are differentiated from fourth-order events by the following criteria:

1. Major episodes of basin filling by thick accumulations of anhydrite, halite, and potash. Assuming little or no post-depositional alteration (i.e., dissolution), the thickness of evaporites may reflect duration of chemical precipitation. Basinal evaporites onlapping the carbonate platform reveal little evidence of wholesale dissolution. Corrosion surfaces, sediment banding, and subaerial transport are not recognized as significant processes. Thicker beds of evaporites are interpreted to represent major relatively long-lived periods of basin restriction and drawdown.

2. Caliche, silcrete, trapped eolian silt, alveolar fabrics, microkarsting, and silt-filled fissures and cracks are observed at the top of all observed sequences. Exposed carbonate terrains can form paleosol features in a relatively short period of time, 100s to a few thousand years (James, 1972; Wright and others, 1988; Rossinsky and Wanless, 1992). Rocks beneath third-order sequence boundaries are subjected to extensive subaerial conditions and show evidence of prolonged exposure when compared to similar rocks deposited beneath interpreted fourth-order sequence boundaries. Platform grainstones under third-order sequence boundaries (e.g., upper Desert Creek and top of Akah) show considerable evidence of subaerial exposure. Meteoric-phreatic diagenesis (leached allochems and recrystallization) extends 25 to 30 feet (8-10 m) beneath sequence boundaries. Meteoric diagenesis below the fourth-order sequence boundaries is much less extensive, and is generally restricted to zones from 0 to perhaps 10 feet (3 m) below sequence boundaries.

3. Regionally extensive black laminated mudstone (e.g., Chimney Rock and Gothic) is observed within transgressive systems tracts overlying third-order sequence boundaries. These mudstones occur over platform highs and are interpreted to have been deposited during major rises in sea level.

4. Significant aggradational growth of phylloid algal banks occurs in early depositional cycles above third-

Figure 25. Recognition of sequence boundaries on the exposed carbonate platform at Honaker Trail, S.E. Utah.

order sequence boundaries. Examples include the lower Desert Creek and lower Ismay phylloid algal banks in the Greater Aneth area. These banks formed from increased accommodation during third-order rises in sea level. Highstand systems tracts, above fourth-order sequence boundaries, are characterized by hydrodynamically controlled grainstone shoals and bars (i.e., oolitic, bioclastic, and peloidal deposits), resulting from shallower water depth and decreased accommodation space. It is important to note that some nearby satellite fields such as the Bug and Papoose Canyon fields are isolated carbonate platforms that formed from biologic activity of phylloid algae in fourth-order sequences. However, these platforms are considerably smaller than the Aneth Platform.

DEPOSITIONAL SYNTHESIS

Since discovery of the Aneth field (1956), 370 million barrels of oil (~1.3 billion barrels original-oil-in-place) have been produced from Middle Pennsylvanian (Desmoinesian) carbonates of the Ismay and Desert Creek intervals. These stratified reservoirs occur within transgressive, highstand, and lowstand systems tracts. Within transgressive systems tracts, dolomudstone/wackestone of the lagoon/tidal flat environment occur within parasequences and display intercrystalline and solution-enhanced secondary porosity. Parasequences of the highstand systems tract represent a time of mound building and platform development as a result of coalescing biologic communities of phylloid algae. Interparticle and shelter porosity dominate. Decreased accommodation within the late highstand resulted in the accumulation of skeletal and nonskeletal wackestone to grainstone. Porosity is developed on paleodepositional highs at the top of parasequences where shoal water facies reveal preserved primary interparticle pore systems. These are secondarily enhanced by leaching of less stable carbonate minerals by meteoric water. In the upper Desert Creek highstand systems tract, oolitic grainstone occurs beneath a 3rd-order sequence boundary and displays oomoldic porosity. In a basin margin position, adjacent to the Aneth platform, hydrocarbons are produced from downslope allochthonous peloidal and oolitic grainstone debris aprons. Siltstone, dolostone, and evaporites form lowstand wedges that were deposited 150 feet (46 m) below the crest of the Aneth carbonate platform. Porous dolomudstone and dolowackestone are productive where they onlap and pinch out against the Aneth carbonate platform.

Nineteen parasequences (i.e., stratigraphic layers or depositional cycles) are described within the Desert Creek and lower Ismay section (gross reservoir section) at McElmo Creek. Each layer top defines the paleodepositional topography, and the consequent depositional geometry helps predict facies within each layer. Geologic maps and cross-sections are constructed to illustrate facies distribution and predict reservoir quality and continuity within each depositional layer.

STRATIGRAPHIC LAYER MODEL

Time-slice mapping of correlatable chronostratigraphic units provides a stratigraphic layer model that is the basis for predicting the distribution and continuity of facies on an interwell and reservoir scale. Systematic vertical and lateral changes in facies stacking in high-frequency parasequences are due to predictable changes in depositional space made available during the relative rising and falling stages of sea level. Recognition of facies and their vertical stacking, the stratigraphic distribution of facies, and reliable time (chronostratigraphic) surfaces are required for prediction of facies within McElmo Creek Unit. In areas uncomplicated by structural/tectonic effects, facies stacking within parasequences is used to develop time lines.

PARASEQUENCES, FACIES STACKING, AND STRATIGRAPHIC LAYERING

As used here, a parasequence is a shoaling upward carbonate cycle that is correlative throughout McElmo Creek Unit (25 sq. mi.; 64 sq. km.) and is interpreted as allocyclic. Interpreted autocycles observed within McElmo Creek are areally restricted accreting tidal flat facies and algal mounds that are nested within parasequences. Typically, autocycles are not correlative from well to well. In the Desert Creek and lower Ismay section, nineteen parasequences or stratigraphic layers are recognized (Table 2). Figure 27 shows the lateral distribution of many of these layers along a platform to basin transect. Fifteen platform layers are recognized in the McElmo Creek area. Four lowstand layers occur in the basin and onlap against the carbonate platform (only three lowstand layers are illustrated in Figure 27; the fourth lowstand cycle [LSW B2] is not present in the northern portion of MCU). Each layer is made up of one or more facies which display evidence of upward shallowing. Evidence of shoaling in cycles includes: 1) upward increase in grain size and grain-supported texture, 2) recognition of features indicative of subaerial exposure near tops of cycles (e.g., caliche, silcrete, alveolar fabric, subvertical fractures filled with fine-grained siliciclastics), 3) upward decrease in abundance and diversity of normal marine biota, 4) upward increase in high-energy nonskeletal allochems such as peloids and ooids, and 5) upward increase in degree of meteoric leaching and grain dissolution. In layers where two or more facies are present, facies stack into asymmetric shoaling-upward cycles. Facies stacking patterns at McElmo Creek are summarized in Figures 28 and 29. Platform cycles differ considerably from basin cycles (Figure 28). In addition, facies associations differ from geologic layer to layer.

GEOLOGY OF MCELMO CREEK UNIT

Vertical and lateral facies changes are summarized here for a few of the significant reservoir layers at McElmo Creek. Depositional structure and facies maps and stratigraphic cross-sections (e.g., Figure 30, oversize, see note at back) illustrate distribution and geometry of facies. The top of Layer IIIC (Figure 27) is selected as the stratigraphic datum because 1) it is interpreted as a low-relief tidal flat surface of essentially 0 meters water depth and 2) a sufficient number of logs penetrates this surface and serves as data points in cross-sections and maps. Successively younger layers are added to the stratigraphic datum. Each layer top is a surface that closely mimics paleodepositional topography.

Qualitative and semi-quantitative estimates of porosity and permeability are discussed where appropriate (for summary of data, see Tables 3 and 4). Porosity values are based on calculated log porosities and results of standard whole core analysis. Estimates of permeability are based primarily on whole core analysis data.

Layer IIE.—

Layer IIE is the first parasequence occurring above the mfs of the Lower Desert Creek sequence and is the initial highstand parasequence of the DS 30 composite sequence (Figure 27). Layer IIE is about 21 feet (6 m) thick and is composed of facies dominated by phylloid algae. An ideal shoaling cycle from base to top consists of: phylloid algal bafflestone (AB), phylloid algal packstone/grainstone (AP/G), and phylloid algal wackestone (AW). This cycle may be capped by thin dolomudstone (DM/W) or dolomitized peloidal grainstone (PP). The phylloid algal bafflestone was deposited in a "deeper" subtidal setting and shows little evidence of meteoric diagenesis. Millimeter-scale shelter porosity is recognized as the most significant pore type. Interestingly, most pores are partially filled with marine fibrous cement. This suggests that large volumes

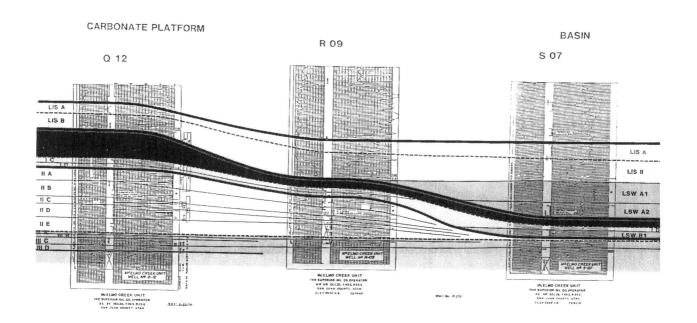

Figure 27. Platform to basin cross-section within the northern portion of McElmo Creek showing stratigraphic layering.

LAYER NUMBER	CHRONOSTRATIGRAPHIC (TIME/ROCK) NAME	LAYER NAME	DISCONTINUITY SURFACE	RANGE OF THICKNESS (FT)
			----4SB	
1.	DESMOINESIAN 40 PS4	LIS HST LIS A		2-38
			----MFS	
2.	DESMOINESIAN 40 PS3	LIS TST LIS B		13-68
3.	DESMOINESIAN 40 PS2	LIS LST LSW A1		0-48
4.	DESMOINESIAN 40 PS1	LIS LST LSW A2		0-39
			3SB---DS40	
5.	DESMOINESIAN 30 PS17	UDC HST IA		2-83
6.	DESMOINESIAN 30 PS16	UDC HST IB		1-44
			----MFS	
7.	DESMOINESIAN 30 PS15	UDC TST IC		0-36
8.	DESMOINESIAN 30 PS14	UDC TST ID		2-32
9.	DESMOINESIAN 30 PS13	UDC LST LSW B1		0-24
10.	DESMOINESIAN 30 PS12	UDC LST LSW B2		0-31
			----4SB	
11.	DESMOINESIAN 30 PS11	LDC HST IIA		0-35
12.	DESMOINESIAN 30 PS10	LDC HST IIB		0-29
13.	DESMOINESIAN 30 PS9	LDC HST IIC		0-32
14.	DESMOINESIAN 30 PS8	LDC HST IID		0-35
15.	DESMOINESIAN 30 PS7	LDC HST IIE		2-32
			----MFS	
16.	DESMOINESIAN 30 PS6	LDC TST IIIA		0-18
17.	DESMOINESIAN 30 PS5	LDC TST IIIB		0-13
18.	DESMOINESIAN 30 PS4	LDC TST IIIC		7-27
19.	DESMOINESIAN 30 PS3	LDC TST IIID		25-42

DESMOINESIAN 40 & 30 - identifies two of the 3rd-order composite sequences in the Desmoinesian section. DS 40 is a third-order sequence boundary and defines the base of DS 40, a third-order composite sequence.

PS = Parasequence
LIS = Lower Ismay
UDC = Upper Desert Creek
LDC = Lower Desert Creek
HST = Highstand systems tract
TST = Transgressive systems tract
LST = Lowstand systems tract
LSW = Lowstand wedge
4SB = 4th-order sequence boundary
3SB = 3rd-order sequence boundary
MFS = Maximum flooding surface

Table 2.   Nineteen stratigraphic layers recognized in the reservoir section at McElmo Creek.

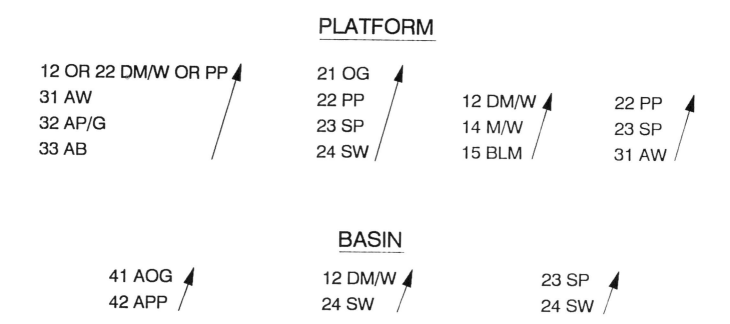

Figure 28. Typical platform and basin facies stacking patterns. Alpha-numeric facies codes are used. See facies descriptions in legend of columnar sections (i.e., Figures 17, 18, and 19) for unabridged facies names.

of marine water were pumped through sediments during and very shortly after deposition. Permeability is generally less than 50 md.

Approximately 10 feet (3 m) of phylloid algal grainstone occurs between the algal bafflestone below and a thin algal wackestone above. The grainstone is interpreted to have been deposited in a beach or nearshore high-energy environment. Algal fronds are fragmented, abraded, and well-sorted. Porosity and permeability are generally favorable (6-10% and 1-500 md). Leaching of algal plates is prevalent along paleodepositional highs (Figure 31). Data indicates that storage and flow capacity are improved in these areas. The parasequence is capped by thin phylloid algal wackestone, dolomudstone, and/or peloidal grainstone. Dolomudstone and dolomitized peloidal grainstone are observed along paleodepositional highs and are interpreted to represent isolated island complexes. On average, permeability is much lower in the capping facies.

Layer IID.—

The compositional make up of Layer IID is very similar to Layer IIE except for the lack of phylloid algal bafflestone (AB). Layer IID ranges from 0 to 0-35 feet (11 m) in thickness. In the basin, some areas underwent little or no sedimentation. Skeletal wackestone is the deeper water basinal equivalent to shallow water carbonate sedimentation on the platform. On the platform, Layer IID is composed of algal grainstone (AP/G) which is overlain by partially dolomitized algal wackestone (AW), peloidal grainstone (PP), and/or dolomudstone (DM/W). Meteoric diagenesis is responsible for the development of algal moldic and isolated vugular porosity. Blocky calcite cement partially fills primary and secondary pores toward the top of the cycle. On the platform, down dip areas are characterized by increased pore filling by blocky calcite cement. The best permeability is developed in phylloid algal grainstone intervals which are dominated by inter-particle porosity. Moldic porosity does not contribute significantly to flow, but does add to storage capacity. As with layer IIE, the best performance probably occurs along subtle paleodepositional highs (i.e., grainstone thicks) at the top of Layer IID (Figure 32). It is in these areas that meteoric diagenetic overprint adds to storage capacity.

Layer IA.—

Layer IA is the uppermost parasequence of the Upper Desert Creek sequence and is capped by the third-order DS 40 sequence boundary (Figure 27). Layer IA is from 12 to 30 feet (4-9 m) thick on the platform and is made up of oolitic grainstone that is subdivided into a stabilized ooid sand flat and a mobile ooid sand belt. Ooids formed along the crest of the platform and were redistributed to other parts of the platform, slope, and basin. On the platform, ooid grain production ceased temporarily midway through Layer IA. Stabilization and bioturbation homogenized the ooid-rich sediment. Shortly after stabilization, carbonate production resumed and an active sand belt became re-established.

Cross-stratified bed forms, coarser grain size, and active marine cementation characterize the upper portion of Layer IA. Early marine cementation is abundant off paleodepositional highs and decreases over these highs. Near the top of the oolitic grainstone, silcretes, calcretes, and caliche pisolites are present and indicate a period of subaerial exposure and vadose diagenesis. Moldic porosity

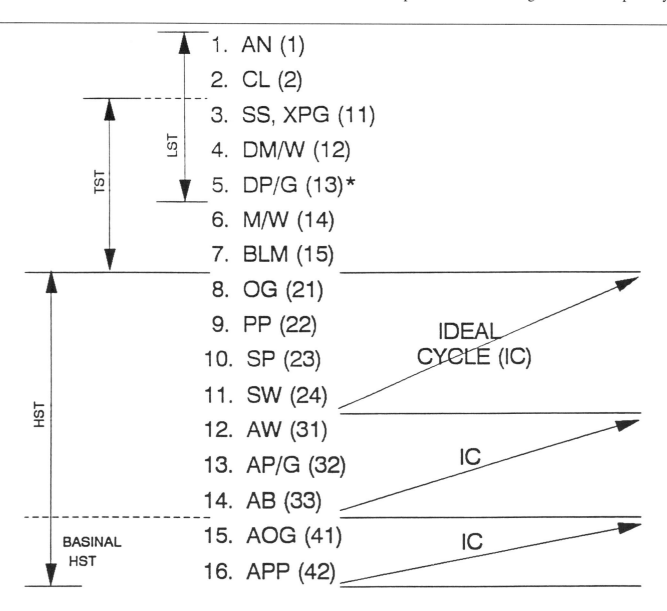

* DP/G (13) also occurs in the highstand systems tract as an allochthonous grainstone deposit.

Figure 29. Ideal cycle stacking at McElmo Creek. Facies are categorized into broadly defined systems tracts. See facies descriptions in legend of columnar sections (i.e., Figures 17, 18, and 19) for unabridged facies designations.

is abundant but poorly connected. Blocky calcite increases away from paleohighs. Porosity varies from 10 to 30 percent, but permeability is usually less than 10 md. Along paleodepositional highs, porosity and permeability improve (Figure 33) because of decreased pore-filling cement.

Figure 33 shows several depocenters (identified as finely diagonal lines) located along the margin of the platform and extending out into the basin. In these areas ooids accumulated as prograding sand sheets on the platform or as carbonate sand aprons along the basin margin. Allochthonous grainstones can exceed 80 feet (24 m) in thickness along the basin margin and extend into the basin for up to several miles as thin sand sheets. In the basin allochthonous oolitic grainstones onlap and abut the platform. Interparticle porosity is the dominant pore-type and there is a direct relationship between grain size and permeability. Coarser ooid grainstones have the best permeabilities (~10 md).

Along the southeastern margin of the platform (horizontal lines in Figure 33), ooid accumulations are thin. Ooids were swept off the platform during high-energy events and sourced thin ooid sand sheets found in the basin. In the western portion of the McElmo Creek Unit (wide-spaced diagonal lines in Figure 33), oolitic grainstone occurs in platform interior areas as thin sand sheets sourced from active sand bars to the east. Sand sheets are thin and exhibit low permeability.

Layer LISA.—

Layer LISA is interpreted to represent the highstand parasequence of the Lower Ismay sequence (Figure 27). Layer LISA averages approximately 24 feet (7 m) in thickness. Algal wackestone (AW) is overlain by skeletal and peloidal packstone and grainstone (SP and PP). Algal wackestone is best developed in subtle lows on the platform. This generally corresponds to a fringe around the pre-existing platform margin. Algal wackestone exhibits porosity values that range from 3 to 12 %. Permeability values range from several millidarcies to several hundred millidarcies. The primary pore network (shelter and interparticle porosity) is enhanced by meteoric leaching of unstable carbonate minerals (i.e., aragonitic algal fronds). Phylloid algae is poorly developed in the southern portion of McElmo Creek and has a patchy distribution in the platform interior (Figure 34). In areas where algae do not proliferate, skeletal wackestone and packstone are time equivalent facies. On paleodepositional highs, Layer LISA is capped by skeletal and peloidal grainstone (PP) (Figure 35). Porosity and permeability is well-developed over these highs.

DEPOSITIONAL AND DIAGENETIC SUMMARY

Three fourth-order depositional sequences are recognized in the Desert Creek and lower Ismay reservoir section at McElmo Creek. The Desert Creek and lower Ismay sequences are interpreted to be type 1 sequences. They are bounded by type 1 sequence boundaries (Vail, 1987) that formed when sea level fell below the Aneth Platform. Each depositional sequence reveals facies that were deposited during lowstand, transgressive, and highstand depositional systems tracts. During relative lowstands of sea level, platform areas were emergent. Sedimentation was confined to basinal lows adjacent to the platforms (Figure 36).

Carbonate platforms formed during initial highstand conditions in the lower Desert Creek as a result of coalescing biologic communities of phylloid algae. Biologic activity resulted in platform aggradation and topographic relief. Interplatform lows received little or no sediment during highstands of sea level. Two types of highstand systems tracts are observed in the Greater Aneth area. Biologically-dominated highstand systems tracts result during third-order sea level rise (e.g., lower Desert Creek and lower Ismay) and form aggradational margins. Progradation is recognized in hydrodynamically-controlled fourth-order highstand systems tracts (e.g., upper Desert Creek). Skeletal, peloidal, and oolitic grainstone dominate shoal water facies in fourth-order HST's (Figure 37).

LOWER DESERT CREEK LOWSTAND SYSTEMS TRACT (LDC-LST)

The LDC-LST occurs beneath the reservoir section in the Greater Aneth area. Lithostratigraphically, the LDC-LST occurs beneath the Chimney Rock and is referred to as the Akah. It is composed of black dolomitic shale, shaly dolomite, sapropelic dolostone, cryptalgalaminite (CL), dolomudstone/wackestone (DM/W), and anhydrite (AN). Regional correlation indicates that the LDC-LST is laterally extensive, forming a thick lowstand wedge deposit (> 150 feet ;>48 m)(Peterson, 1966) that onlaps against the Akah platform (see Peterson, 1966; Goldhammer and others, 1991). Several deep wells penetrate this interval in the Greater Aneth area, and gas shows are indicated on mud logs.

FACIES	MEAN ø (%)	RANGE ø (%)	ARITH. MEAN K (md)	RANGE K (md)	GEOMETRIC MEAN K (md)	COUNT K (md)	REMARKS
AN	1	0-5	0.04	0.03-0.07	0.04	5	Basin Facies
CL	4	2-13	0.01	0.00-0.29	0.00	24	Basin Facies
SS	3	0-17	0.15	0.00-2.14	0.01	34	Mostly Basin
DM/W	9	0-28	2.59	0.00-66	0.18	237	Platform and Basin
DP/G	5	1-16	2.58	0.02-17	0.29	14	Platform Margin and Basin
M/W	3	0-15	0.47	0.00-7.37	0.04	34	Platform and Basin
BLM	-	-	-	-	-	-	
MOG	14	1-29	6.50	0.00-165	0.36	58	Platform
SOG	14	2-30	1.27	0.02-8.80	0.37	52	Platform
OG	10	2-24	6.06	0.00-67	0.22	27	Undifferentiated SOG and MOG
PP	6	0-19	2.10	0.00-67	0.11	111	Platform
SP	3	0-12	0.31	0.00-6.34	0.02	81	ø, k better on Platform; fracture k?
SW	3	0-13	0.08	0.00-1.15	0.01	56	Platform and Basin
AW	5	0-17	2.66	0.00-106	0.143	228	
AP/G	7	1-14	51.37	0.00-1000	3.201	118	
AB	6	4-10	40.08	0.02-727	1.24	21	
APP	1	0-2	0.84	0.00-6.69	0.03	17	Basin

Numbers in table are based on results of standard core analysis data for the following wells: V-08, V-06, U-07, U-05, T-08, T-06, T-04, S-08, S-07, S-05, S-03, R-19, R-08, R-06, Q-16, Q-12, Q-10, Q-08, O-16 and J-131. Results of AOG facies are not reported here but numbers should be similar to that of DP/G facies. Permeability greater than 1 darcy is excluded from this summary.

Table 3. Mean and range of porosity and permeability for facies observed at McElmo Creek. Alpha-numeric facies codes are used. See facies descriptions in legend of columnar sections (i.e., Figures 17, 18, and 19) for unabridged facies names.

	PLATFORM					BASIN				
LAYER	MEAN Ø (%)	RANGE Ø (%)	MEAN K (md)	RANGE K (md)	COUNT	MEAN Ø (%)	RANGE Ø (%)	MEAN K (md)	RANGE K (md)	COUNT
LIS A	4	1-17	5.38	0.01-263	78	4	0-10	4.88	0.01-36	28
LIS B	9	1-27	3.03	0.01-66	62	6	0-28	2.05	0.01-23	20
LSW A1	--	--	--	--	--	3	0-13	0.02	0.01-0.29	29
LSW A2	--	--	--	--	--	4	0-13	0.91	0.01-17	46
I A	14	0-30	3.99	0.01-165	111			FEW DATA AVAILABLE		
I B	4	0-14	0.29	0.01-2.5	52			FEW DATA AVAILABLE		
I C	7	1-18	3.07	0.01-86	44	5	0-17	2.36	0.01-24	23
I D	12	1-25	1.72	0.01-10	37	3	0-11	1.04	0.01-10.1	21
LSW B1	--	--	--	--	--	5	1-11	1.02	0.01-4.6	11
LSW B2	--	--	--	--	--			DATA NOT AVAILABLE		
II A	8	0-24	4.31	0.01-67	77	4	0-9	0.80	0.02-6.34	20
II B	7	1-26	13.79	0.01-197	39	1	0-2	0.50	0.01-2.46	13
II C	7	1-14	20.17	0.02-1000	16	11	6-15	1.93	0.05-10	18
II D	7	1-14	88.44	0.01-950	26	--	--	FEW DATA AVAILABLE		
II E	8	1-28	17.58	0.01-727	114	5	0-9	9.75	0.01-137	44
III A	--	--	--	--	--	5	1-15	0.76	0.01-11	24
III B	--	--	--	--	--	15	5-22	0.35	0.01-1.3	12
III C	--	--	--	--	--	8	2-24	0.82	0.01-13	53
III D	--	--	--	--	--	8	3-18	3.48	0.01-53	19

Numbers in table are based on results of standard core analysis data for the following wells: V-08, V-06, U-07, U-05, T-08, T-06, T-04, S-08, S-07, S-05, S-03, R-19, R-08, R-06, Q-16, Q-12, Q-10, Q-08, O-16 and J-131. Results of AOG facies are not reported here but numbers should be similar to that of DP/G facies. Permeability greater than 1 darcy is excluded from this summary.

Table 4.   Mean and range of porosity and permeability by layer for the platform and basin in McElmo Creek.

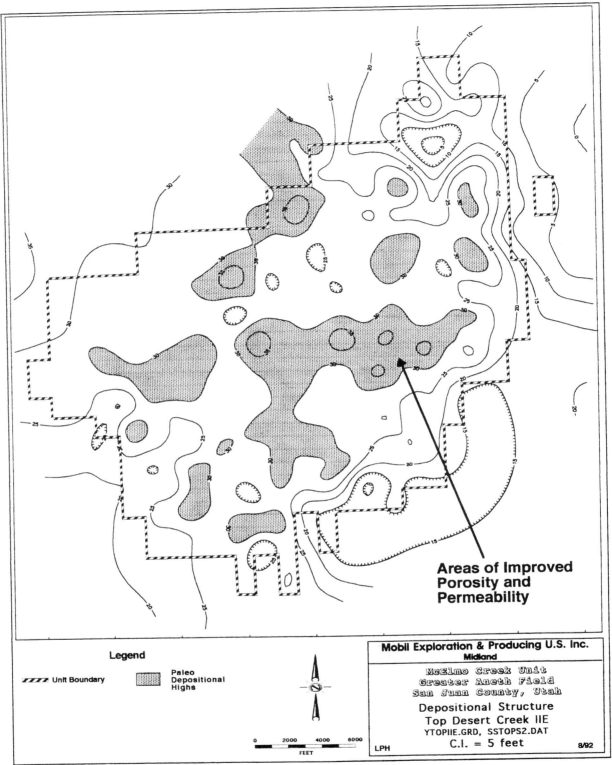

Figure 31. Depositional structure map on top of Layer IIE. Highlighted areas show paleodepositional highs where diagenesis improved storage capacity of the rock. Core analysis data suggests that permeability is improved in many of the paleohighs. Dolomudstone and dolomitized peloidal grainstone occur on some of these highs as isolated beach/island complexes. Contour numbers correspond to number of feet above stratigraphic datum. Areas highlighted are > 30 ft in thickness.

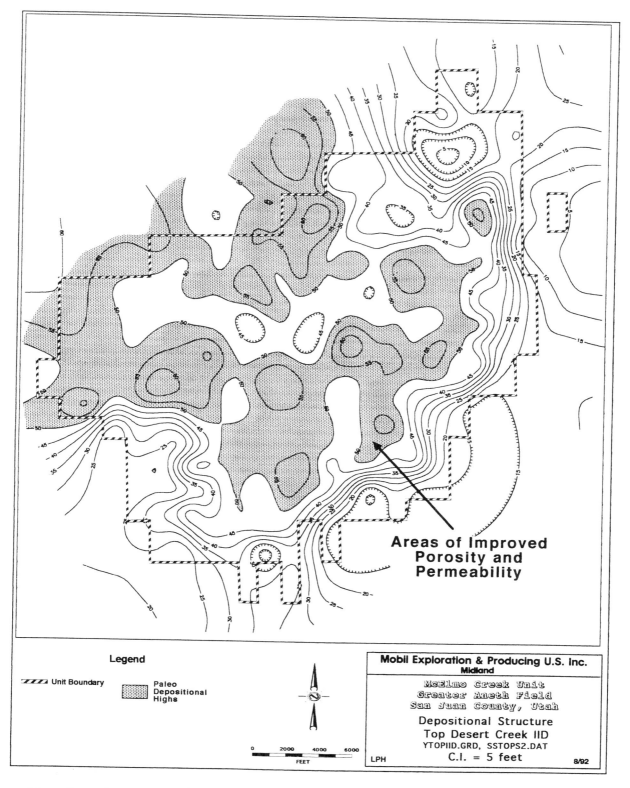

Figure 32. Depositional structure map on top of Layer IID. Highlighted areas show paleodepositional highs where porosity and permeability are expected to be well-developed. Contour numbers correspond to number of feet above stratigraphic datum. Areas highlighted are >50 ft in thickness.

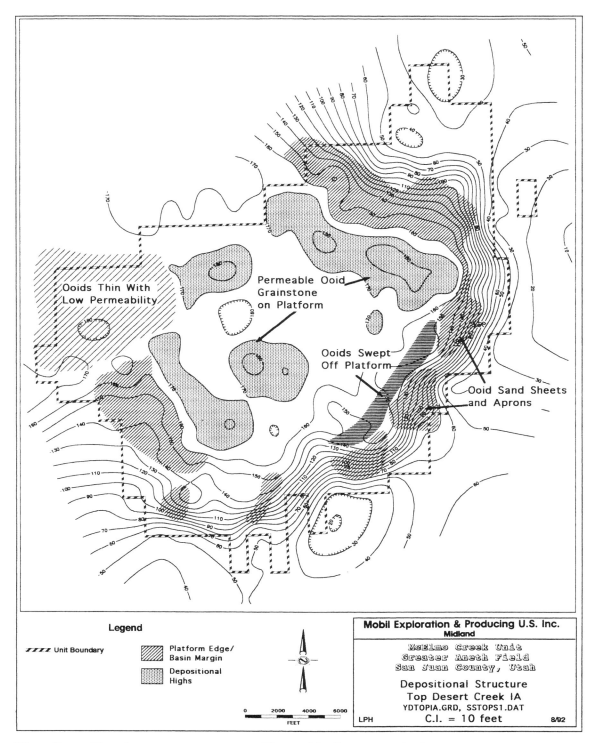

Figure 33. Depositional structure map on top of Layer IA. Dot pattern indicates paleodepositional highs where decreased circulation of sea water resulted in less marine cementation, more preserved interparticle porosity, and greater permeability. Fine diagonal lines show where ooids accumulated in down dip areas as prograding sand sheets and sand aprons. Horizontal lines show areas of thin ooid grainstone deposits. Ooids were swept off the platform during high-energy events and sourced thin ooid sand sheets in the basin. Wide-spaced diagonal lines show platform interior areas where ooid deposits are thin and exhibit low permeability. Contour numbers correspond to number of feet above stratigraphic datum.

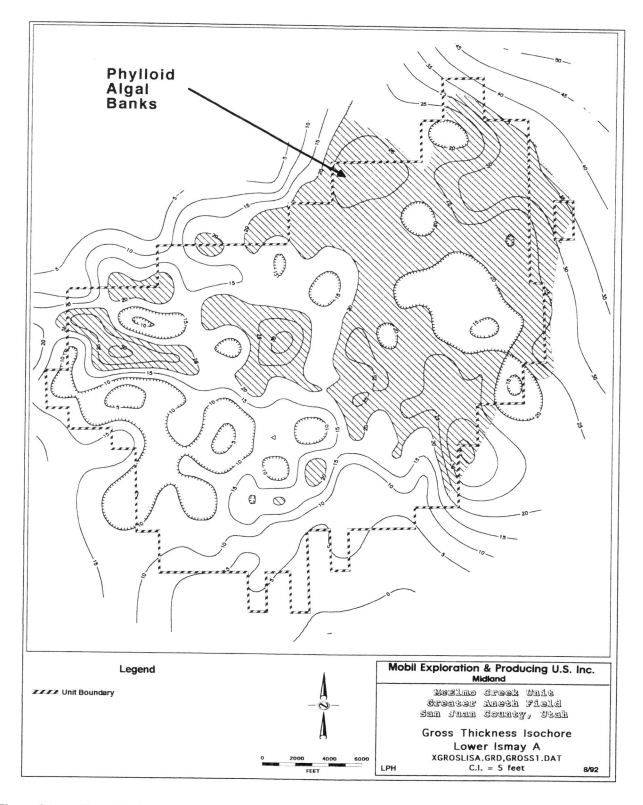

Figure 34. Gross thickness isochore of Layer LISA. Phylloid algal facies are best developed in highlighted areas. Gross thickness values <10 ft indicate areas where phylloid algal facies did not develop.

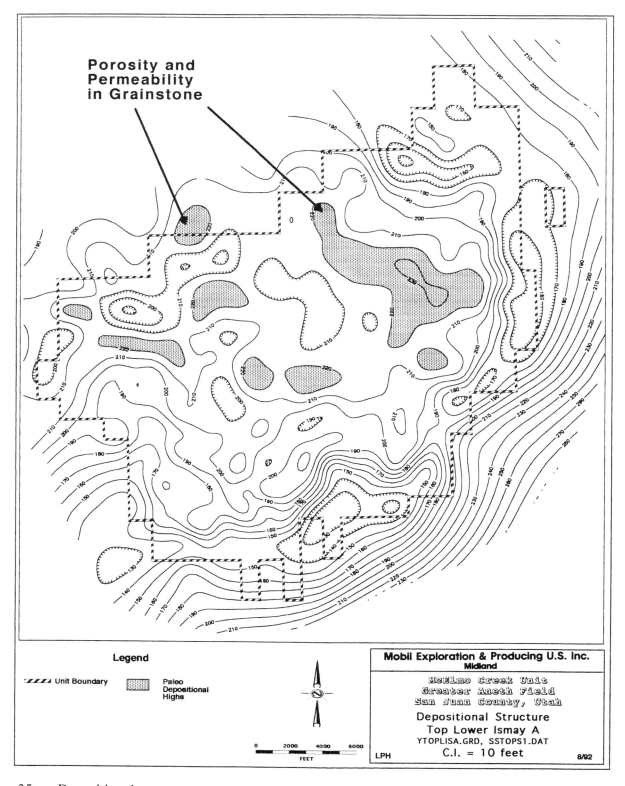

Figure 35. Depositional structure map on top of Layer LISA. Highlighted areas highlighted indicate paleodepositional highs where porous and permeable grainstone cap depositional cycles. Contour numbers correspond to number of feet above stratigraphic datum.

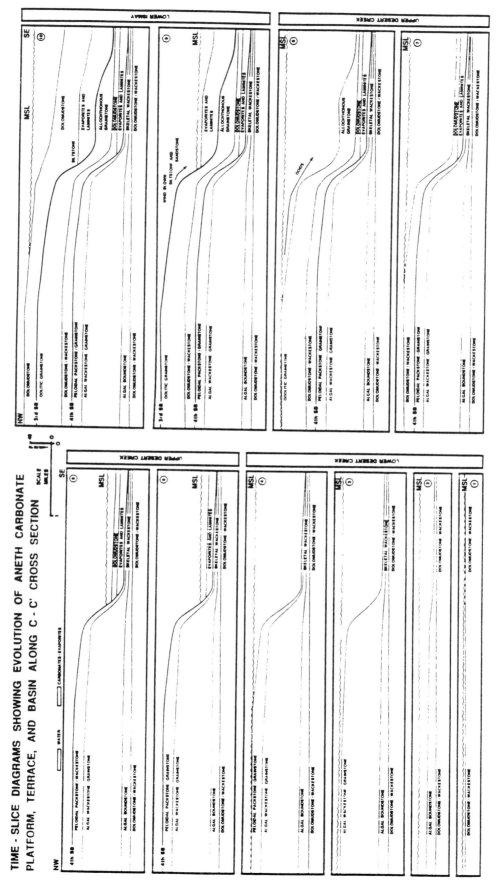

Figure 36. Time-slice diagrams showing the evolution of the Aneth Platform to basin transition. Three depositional sequences are recognized. They are the lower Desert Creek, upper Desert Creek, and lower Ismay. Note facies changes during relative changes in sea level.

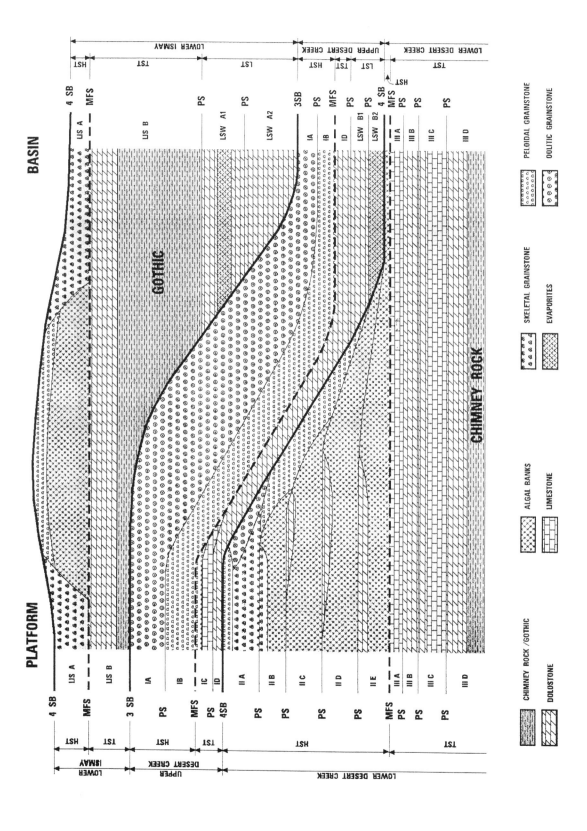

Figure 37. Schematic cross-section of the platform and basin at McElmo Creek.

LOWER DESERT CREEK TRANSGRESSIVE SYSTEMS TRACT (LDC-TST)

The LDC-TST is composed of four laterally continuous low relief parasequences (Layers IIIA-IIID) of fairly uniform thickness (Figure 27). The Chimney Rock (BLM Facies) and the underlying skeletal wackestone facies forms the base of Layer IIID and is basinwide in distribution. The Chimney Rock marks the onset of regional transgression into the salt depocenter of the Paradox basin. Concomitant dissolution of soluble salts (i.e., halite, sylvite, etc.) raised oceanic salinities in the Paradox basin and facilitated cyanobacterial blooms and preservation of organic matter. The Chimney Rock is thicker in the Greater Aneth area than in surrounding areas (Peterson, 1966). Increased thickness may represent sediment accumulation in a subtle depression.

Each parasequence within the LDC-TST was deposited in shallow marine water and is capped by subaerially exposed phylloid algal wackestone (AW; Layer IIIA) or tidal flat dolostone (DM/W; Layers IIIB, C, and D). Vugular and algal moldic porosity in Layer IIIA is occluded with blocky calcite cement. Porosity is developed in the dolostone, but permeability is sporadic. Reduction of permeability results from 1) disseminated fine-grained quartz silt occurring in what would otherwise be intercrystalline pore space (Layers IIIB and IIID) and 2) partial filling of intercrystalline porosity by cement (i.e., calcite, anhydrite, and/or silica).

In the LDC-TST the best reservoir is developed in Layer IIIC where phylloid algal wackestone (AW) is overlain by dolomudstone (DM/W). The layer is generally less than 20 feet (6 m) thick. Porosity and permeability is best developed in the central portion of MCU where this unit fills subtle topographic lows. It is in this area where the AW Facies is a more significant contributor. The AW Facies is dominated by moldic (algal moldic) porosity, whereas, the DM/W Facies is dominated by intercrystalline porosity. Within Layer IIIC, permeability is less than 13 md and averages 1 md (Table 4).

Along the eastern and southern margin of McElmo Creek Unit, layers IIIA and IIIB are thicker and were deposited on paleodepositional highs as tidal flats. The tidal flats are composed of finely crystalline dolomite with well-developed intercrystalline porosity. These dolostones flank a subtle bathymetric low (central MCU), an area that later became the Aneth Platform. Layers IIIA (AW) and IIIB (DM/W) are deposited in the bathymetric lows and are barriers to vertical fluid flow throughout much of MCU. Low permeability values are reported for Layers IIIA and IIIB (Table 4).

LOWER DESERT CREEK HIGHSTAND SYSTEMS TRACT (LDC-HST)

Five shallowing-upward depositional cycles (Layers IIA-IIE) characterize the LDC-HST (Figure 27). Cycle caps reveal subaerial exposure, but duration and areal extent of exposure was limited. Subaerial exposure is indicated by the development of localized partially dolomitized grainstone shoals and tidal flat mudstones that cap parasequences. In general, porosity and permeability are developed best in areas of subaerial exposure where meteoric diagenesis resulted in dissolution of unstable carbonate allochems. In these areas, the storage capacity of the rock has increased, and permeability is enhanced within thin stringers. High permeability thief zones occur toward the top of depositional cycles. They direct injection fluids along conduits and reduce injection into and production out of adjacent strata.

The Aneth Platform formed during relative sea level rise in the LDC-HST. Biologic communities composed dominantly of phylloid algae formed discrete mounds that in time coalesced into extensive algal banks that kept pace with relative sea level rise. Phylloid algae are the most important constituents on the platform within Layers IIC-IIE. These parasequences stack and form an aggradational platform margin. The platform margin contracts as each successive layer is added (Figures 38, 39, and 40).

The base of Layer IIE coincides with the maximum flooding surface (mfs). Just above the mfs, 10 to 12 feet (3-4 m) of algal bafflestone (AB) formed in subtle bathymetric lows. The AB Facies was deposited just after rapid sea level rise in a catch-up phase of highstand sedimentation. Deposited algal sheaths drape over one another and are cemented early by marine fibrous and botryoidal cement forming a rigid "reef" framework. A complex network of primary shelter and interparticle pores are the dominant pore types. Pores are partly to completely filled with marine cement as a result of active fluid movement through the primary pore network. Algal grainstone (AP/G) and wackestone (AW) overlie algal boundstone (AB) and cap the parasequence (Layer IIE). Meteoric diagenesis played a minimal role in altering the AB Facies.

Layers IIA-IID represent a shoal water or keep-up style of highstand sedimentation. In these layers meteoric diagenesis played an important role in modifying storage capacity of the rock. Isolated grainstone shoals and

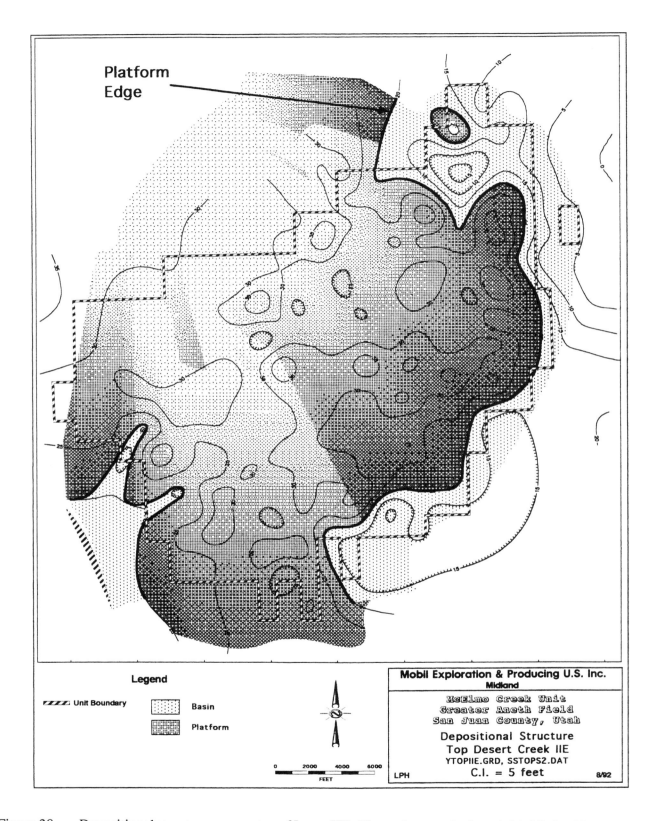

Figure 38. Depositional structure map on top of Layer IIE. The carbonate platform is highlighted in cross-hatch pattern; the basin in dot pattern.

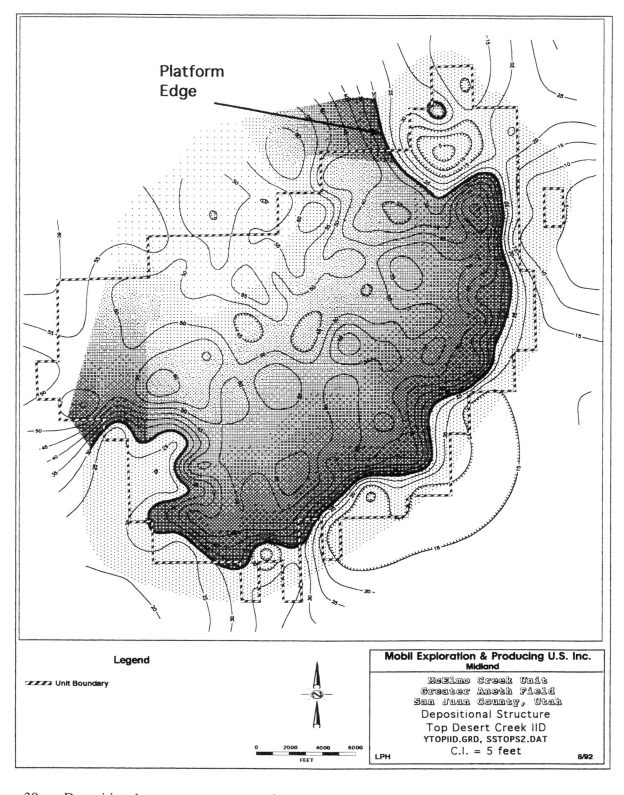

Figure 39. Depositional structure map on top of Layer IID. The carbonate platform is highlighted in cross-hatch pattern; the basin in dot pattern.

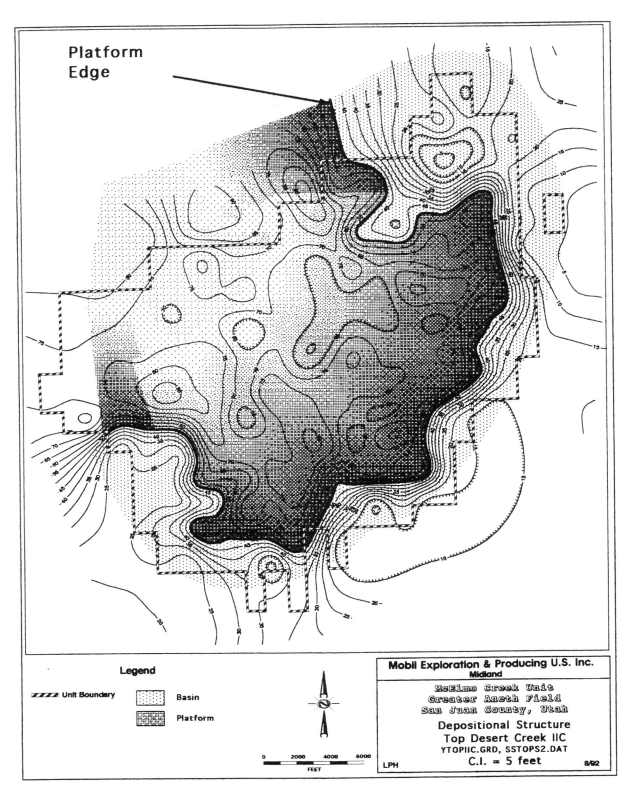

Figure 40. Depositional structure map on top of Layer IIC. The carbonate platform is highlighted in cross-hatch pattern; the basin in dot pattern.

emergent islands formed on the platform. An exception to this style of sedimentation occurs along the northern, windward margin of the platform, where the last major occurrence of phylloid algae is recognized in Layer IIB in agitated and slightly deeper water.

Layers IIA and IIB are composed of skeletal, oolitic, and peloidal wackestone to grainstone that formed during late highstand on a shallow water platform. In the basin, layers IIA-IIE are composed primarily of skeletal wackestone and/or packstone. These deposits are thin, generally less than 10 to 20 feet (3-6 m), whereas, on the platform, time equivalent rocks exceed 100 feet (31 m) in thickness. In Layer IIA peloidal grainstone is widely distributed on the platform and occurs 30 to 65 feet (10-20 m) below the crest of the platform. Peloidal grainstone is interpreted to represent a shallow water facies and is not likely to have formed along a platform margin setting in a range of 3 feet (1 m) to 30-67 feet (10-20 m) of water depth. This suggests that either 1) peloidal grainstone formed on paleodepositional highs and was redistributed to other platform areas as allochthonous debris and/or 2) peloidal grainstone was deposited in shoal water during late highstand as sea level fell resulting in an outbuilding of essentially autochthonous grainstone. The absence of allochthonous peloidal grainstone in the basin suggests that late highstand outbuilding may have been operative. In addition, along the edge of the platform in the southern part of McElmo Creek, an oolitic grainstone shoal formed on a terrace 50 feet (15 m) below the crest of the platform. At the time of ooid shoal development, most platform areas were subaerially exposed.

UPPER DESERT CREEK LOWSTAND SYSTEMS TRACT (UDC-LST)

Two depositional cycles are identified in the upper Desert Creek Lowstand Systems Tract (Figure 27). Lowstand Wedge B2 (LSWB2) is overlain by LSWB1 and both are composed of dolostone and fine-grained quartz siliciclastics. Evaporites occur in LSWB2, which is areally restricted to the extreme southwestern portion of McElmo Creek. LSWB1 and LSWB2 occur in topographic lows in intrashelf depressions adjacent to the carbonate platform. Both parasequences were deposited during a regional lowstand of sea level and thin laterally by onlap against paleodepositional highs. Shallow subaqueous sedimentation was punctuated by periodic subaerial exposure. Porous and permeable dolostone reservoirs occur within the UDC-LST. Although they may onlap porous algal facies on the platform, lower permeability in the dolostone should preclude significant cross-flow unless mechanical stimulation creates conduits of fluid flow.

While sedimentation occurred in the basin, extended conditions of subaerial exposure modified porosity and permeability on the carbonate platform. Paleodepositional highs and isofacies maps showing grainstone thicks should be targets for pattern analysis and production optimization. However, it appears that paleohighs on the platform were most effected by vadose diagenesis. Calcretes and silcretes formed in some of these areas and reduced vertical and possibly horizontal permeability within the uppermost 3 to 6 feet (1-2 m) of the PP Facies.

UPPER DESERT CREEK TRANSGRESSIVE SYSTEMS TRACT (UDC-TST)

The UDC-TST is composed of two parasequences (Layers IC and ID) that backstepped during regional sea level rise (Figure 27). The dolostone/wackestone Facies (DM/W) was deposited in shallow water in the basin, above the lowstand wedge deposits. As sea level rose, sedimentation ceased in the basin. Upon submergence of the platform, DM/W Facies was deposited regionally as a shallow water peritidal accumulation that draped the carbonate platform. Layer ID was deposited in shallow water in the basin during early transgression and on the platform during later stages of transgression.

Layer IC is absent in the basin or is represented as a thin dolomitic mudstone which overlies the dolomudstone (DM/W) of Layer ID. Layer IC is overlain by highstand deposits of diverse limestone lithologies. On the platform Layer IC is a shallowing-upward cycle that is capped by locally emergent skeletal and peloidal grainstone. Along the northern and southern margins of the platform, oolitic grainstone banks accumulated in shallow water.

UPPER DESERT CREEK HIGHSTAND SYSTEMS TRACT (UDC-HST)

Two high-frequency depositional cycles (Layers IA and IB) are recognized in the UDC-HST (Figure 27). Above the maximum flooding surface (MFS), 5 to 25 feet (2-8 m) of skeletal wackestone and lesser packstone overlie peloidal and skeletal grainstone, which cap the uppermost parasequence of the UDC-TST. The MFS is a sharp boundary that separates shallow water platform sedimentation below from deeper water, but still relatively shallow mud-rich skeletal facies above. Skeletal wackestone is recognized above the MFS throughout the MCU area and is probably a barrier to vertical fluid movement. On the platform Layer IB is capped by peloidal packstone/ grainstone (PP); subaerial exposure is observed in isolated areas. In the basin Layer IB is made up of skeletal wackestone which is overlain by fine-grained allochthonous peloidal grainstone.

The latter part of the UDC-HST is characterized by deposition of oolitic grainstone (Layer IA) in shallow, agitated, normal marine water. Ooids formed on the platform and were redistributed on the platform and into the basin. Highstand shedding of ooids, peloids, and skeletal debris from the platform to the basin is suggested as the dominant mechanism to explain the occurrence of shallow-water-derived allochems in basinal settings. Four lines of evidence are used to support this hypothesis.

1. The same succession of grain types is observed on the platform and in the basin. On the platform the UDC-HST is composed of skeletal wackestone/packstone, which is overlain by peloidal grainstone. The platform succession is capped by oolitic grainstone. In the basin skeletal wackestone is overlain by peloidal/skeletal packstone/ grainstone, which is overlain by oolitic grainstone. These basinal deposits occur 120 to 140 feet (37-43 m) below the crest of the platform.

2. The allochthonous grainstone deposits are composed primarily of limestone just as their time-equivalent deposits on the platform. In the basin, allochthonous limestone debris is bounded by dolostone, siltstone, and anhydrite, indicative of lowstand and early transgressive conditions. It should be noted that near the sequence boundary, dolomitization may alter the uppermost 3 to 6 feet (1-2 m) of allochthonous grainstone. Dolomitization probably results from downward seepage of dolomitizing fluids that altered the overlying lowstand deposits.

3. Platform ooids were subjected to meteoric diagenesis and display oomoldic porosity. Allochthonous grainstone deposits contain ooids with well-preserved ooid cortices. Subaerial exposure and meteoric leaching of ooid cortices did not occur in the basin.

4. A sharp corrosional and mineralized contact separates allochthonous oolitic grainstone from overlying dolostones and evaporites of the lower Ismay Lowstand Systems Tract (LIS-LST).

LOWER ISMAY LOWSTAND SYSTEMS TRACT (LIS-LST)

Two depositional cycles are recognized in the basin. Lowstand wedges (LSWA1 and LSWA2) are composed of shallow subaqueous facies (DM/W, DP/G, AN, and CL) that are restricted to topographic lows (Figure 27). These facies onlap against the carbonate platform. Relative fall of sea level exposed the carbonate platform to vadose and phreatic meteoric diagenesis. Carbonate grains were cemented and leached soon after exposure. In the northern portion of MCU, a basal transgressive siltstone sharply overlies platform carbonate and contains within it reworked cobbles of lithified oolitic grainstone with oomoldic porosity.

Porous platform margin deposits (layers IA through IIE) underwent dolomitization during the regional sea level fall and lowstand. Restricted basinal brines are interpreted to have interacted with porous rock along the margin of MCU and caused pervasive dolomitization (Figure 41). Dolomitization resulted in enhanced porosity and permeability. This is especially marked along the windward northern margin of the platform where storms and other higher energy events are interpreted to have pushed heavier brines onto terraces, where refluxing brines percolated down through the platform margin deposits.

LOWER ISMAY TRANSGRESSIVE SYSTEMS TRACT (LIS-TST)

Relative rise in sea level resulted in the deposition of Layer LISB (Figure 27). Regionally correlative skeletal wackestone marked the onset of transgression. The Gothic BLM was deposited next as chemical conditions similar to those interpreted during Chimney Rock BLM deposition occurred over platform and basin areas. Continued sedimentation in the salt basin prevented dissolution of salt and water chemistry soon returned to normal conditions. The LISB parasequence is capped by a thick tidal flat in some areas of McElmo Creek. The tidal flat environment underwent subaerial exposure and dolomitization. Today, these dolostones exhibit reservoir quality porosity and permeability, averaging 15% and 3 md., respectively. Permeability varies considerably and exceeds 50 md. within stratified centimeter-scale intervals. Laterally adjacent to tidal flats are paleodepositional lows which are filled with thin mudstone and wackestone (M/W).

LOWER ISMAY HIGHSTAND SYSTEM TRACT (LIS-HST)

The lower Ismay Highstand Systems Tract is made up of the parasequence Layer LISA (Figure 27). The mfs is a sharp surface and separates dolostone (DM/W) below from phylloid algal wackestone (AW) above. Algal mound development was less extensive in the lower Ismay than in the lower Desert Creek. Thickest banks are less than 33 feet (10 m) thick and occur along the platform margin in areas where sufficient depositional space was available for optimum growth of phylloid algae. Crestal areas of the platform reveal thin patch banks or skeletal packstone/wackestone. LISA is capped by subaerially exposed skeletal and peloidal packstone and grainstone. Above the sequence boundary, lowstand wedge evaporite deposits of

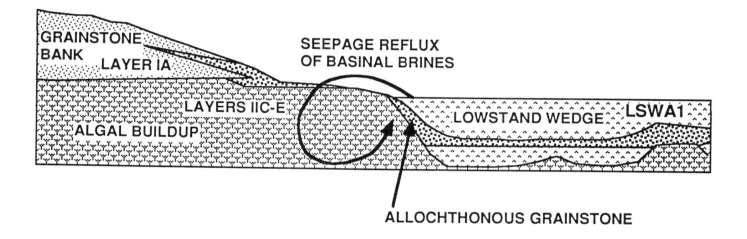

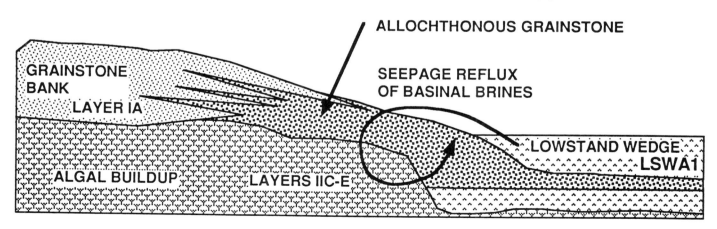

Figure 41. Two platform to basin cross-sections (northwestern and southeastern MCU) showing relationship between basinal fluids associated with the lower Ismay Lowstand Systems Tract and dolomitization of platform margin deposits.

the upper Ismay accumulated in the southern portion of MCU.

RESERVOIR QUALITY

Reservoir quality is based primarily on thickness, lateral distribution, porosity, and permeability of the reservoir facies present within McElmo Creek. Porosity values are based on calculated log porosities and results of standard whole core analysis (Tables 3 and 4). Estimates of permeability are based primarily on whole core analysis data (Tables 3 and 4).

Fortunately, within the Middle Pennsylvanian carbonates at McElmo Creek, diagenesis does not significantly cross-cut depositional fabric. As a result, facies are identified and arranged into shallowing-upward packages, which ultimately are placed in a parasequence framework. The framework allows us to conduct high resolution mapping to predict facies distribution and geometry. An understanding of porosity and permeability within facies enables us to identify stratified reservoirs in carbonate rocks.

Three types of porosity are observed in the Desert Creek and lower Ismay cyclic carbonates at McElmo Creek: 1) intercrystalline and solution-enhanced secondary porosity occur in dolostone of tidal flat origin; 2) solution-enhanced primary porosity (interparticle and shelter with lesser moldic porosity) is present in nearshore, beach, and buildup facies; and 3) abundant secondary porosity results from meteoric solution of unstable carbonate minerals in shallow subtidal and peritidal facies that cap shallowing-upward cycles.

Intercrystalline and solution pores in dolomite rocks are very small (0.025-0.25 mm across). In the highstand systems tracts, dolostone is thin and laterally discontinuous, but porosity and permeability are well-developed locally.

Thickest dolostone reservoirs occur as lowstand wedges in lowstand systems tracts. The Dolomudstone/Wackestone Facies (DM/W) is laterally continuous around the edge of the platform. However, porosity and permeability are developed locally in isolated reservoirs because anhydrite and calcite fill pore space and reduce permeability significantly. Although tidal flat dolostone reveals good log porosity, permeability is sporadically developed and production has shown that these rocks are poor to excellent oil reservoirs. The best reservoirs are expected in TST's on the carbonate platform in ID and at the top of IIIC and LISB. Pore systems are laterally extensive and not filled with anhydrite, late calcite, or baroque dolomite. In 1959 the MCU H-24 was completed as a Desert Creek producer. In 1970 a high water cut forced remedial action. It was decided to perforate the dolostone at the top of LISB. The well flowed 710 BOPD during the initial production test, and over the next 6 months it averaged more than 250 BOPD.

Solution-enhanced primary porosity (interparticle and shelter with lesser moldic porosity) is developed best in highstand systems tracts associated with algal buildups (AP/G, AB, and AW) in layers LISA and IIC-IIE, peloidal grainstone shoals (PP) in Layer IIA, and allochthonous carbonate sand aprons (DP/G and AOG) that occur along the edge of the platform in Layer IA. In general, these reservoirs provide the best long term production and probably account for the majority of oil produced in McElmo Creek. Although permeability varies considerably in these facies, it is more uniformly distributed when compared to reservoirs dominated by facies revealing intercrystalline and moldic pores. In addition, the facies that reveal abundant interparticle porosity tend to be thicker and more continuous laterally. Facies dominated by primary porosity as opposed to secondary (moldic) porosity reveal lower storage capacity per unit volume, but higher permeability.

Moldic porosity occurs throughout the Desert Creek and lower Ismay section, but is particularly common in cycle capping facies in highstand systems tracts. Algal Wackestone (AW), Nonskeletal/Skeletal Packstone/Grainstone (PP), Oolitic Grainstone (OG), and thin dolostone (DM/W) reveal evidence of dissolution.

In general, facies composed dominantly of moldic porosity in the absence of significant primary porosity (interparticle or shelter) are poor reservoirs. They may be quite porous but permeability is low, usually much less than 1 md. Paleodepositional highs beneath sequence boundaries are targets because they are commonly dominated by grainstone facies that reveal interparticle porosity. Intense meteoric diagenesis and the development of moldic porosity added to the storage capacity of the rock, generally without a detrimental effect to permeability. Moldic porosity is developed best on the carbonate platform below sequence boundaries in Layers IIA, IA, and LISA. Most notable is Layer IA, where subaerial exposure and meteoric diagenesis leached 20 to 30 feet (6-9 m) of oolitic grainstone. In these rocks, ooid cortices were dissolved forming oomoldic porosity; interparticle porosity between ooids was occluded by marine fibrous cement. Storage capacity is favorable, but permeability is low. Figure 33 shows isolated paleodepositional highs where marine cementation was less pervasive. In these areas interparticle porosity is preserved and flow capacity of the rock is improved significantly. Some of the best Layer IA reservoirs occur along paleodepositional highs. A recent recompletion in one of the paleodepositional highs, the MCU M-12B, resulted in increased uplift when perforated in Layer IA. The well was completed in 1978 as a producer in a deeper horizon. In 1986 it was shut in because of high water production, but in October of 1990, the Reservoir Management Team recommended completing this well in Layer IA, a shallower reservoir. Initial production averaged 400 BOPD and as recently as March 1992 it was making 100 BOPD.

CONCLUSIONS

The sequence stratigraphic framework constructed for the Middle Pennsylvanian (Desmoinesian) of southeastern Utah has allowed identification of major exposure and flooding surfaces that serve as regional time lines or chronostratigraphic surfaces. Nineteen discrete and mappable high-frequency parasequences were defined and correlated within the Desert Creek and lower Ismay section (Middle Desmoinesian) at the McElmo Creek Unit (MCU) of the Greater Aneth field, SE Utah. Facies stacking within parasequences allowed: 1) mapping and prediction of the distribution of porous and permeable facies; and 2) characterization of the variability in reservoir pore systems.

Three fourth-order depositional sequences are recognized in the Desert Creek and lower Ismay reservoir section at McElmo Creek. Each depositional sequence reveals facies that were deposited during lowstand, transgressive, and highstand depositional systems. During highstands of sea level, sediment accumulation occurred on the carbonate platform; topographic lows adjacent to the platform received little or no sediment. During lowstands of relative sea level, the platform was emergent, and sedimentation took place in topographic lows adjacent to the platform.

Two types of highstand systems tracts are observed in MCU. Biologically-dominated highstand systems tracts result during sea level rises (e.g., lower Desert Creek and lower Ismay) and form aggradational margins. Progradation is recognized in hydrodynamically-controlled highstand systems tracts (e.g., upper Desert Creek).

At McElmo Creek stratified reservoirs occur within transgressive systems tracts (TST), highstand systems tracts (HST), and lowstand systems tracts (LST). Within the transgressive systems tracts, dolomudstone/wackestone of the lagoon/tidal flat environment comprise parasequences and display intercrystalline and solution-enhanced secondary porosity. In the lower Desert Creek (LDC), initial parasequences of the highstand systems tracts (layers IIC-IIE) represent a time of mound building and platform development as a result of coalescing biologic communities of phylloid algae; interparticle and shelter porosity dominate. Subsequent parasequences within the LDC-HST exhibit skeletal and nonskeletal wackestone to grainstone. Porosity is developed on paleodepositional highs at the top of parasequences where shoal water facies possess preserved primary pore systems that are secondarily enhanced by meteoric leaching of less stable carbonate minerals. In the upper Desert Creek HST (UDC-HST), oolitic grainstone occurs beneath a 3rd-order sequence boundary and displays oomoldic porosity. In a basin margin position, adjacent to the Aneth platform, hydrocarbons are produced from downslope allochthonous UDC-HST peloidal and oolitic grainstone debris aprons. Siltstone, dolostone, and evaporites form lowstand wedges that were deposited 150 feet below the crest of the Aneth carbonate platform. Porous dolomudstone and dolowackestone are productive where they onlap and pinch out against the Aneth carbonate platform.

Three types of porosity are observed in the Desert Creek and lower Ismay cyclic carbonates at McElmo Creek:

1. Intercrystalline and solution-enhanced secondary porosity occur in dolostone of tidal flat origin. Intercrystalline and solution pores in dolomite rocks are very small (0.025-0.25 mm across). However, interconnecting pathways exist, especially in dolomudstone (DM/W) deposited on the platform on paleodepositional highs in the transgressive systems tracts. Core analysis and production/performance data indicate that significant fluid pathways are developed on the carbonate platform in parasequence ID and at the top of parasequences IIIC and LISB.

2. Solution-enhanced primary porosity (interparticle and shelter with lesser moldic porosity) is present in nearshore, beach, and buildup facies. Primary porosity is developed best in highstand systems tracts associated with algal buildups (AP/G, AB, and AW) in parasequence layers LISA and IIC-IIE, peloidal grainstone shoals (PP) in parasequence Layer IIA, and allochthonous carbonate sand aprons (DP/G and AOG) that occur along the edge of

the platform in parasequence Layer IA. These reservoirs provide the best long term production and probably account for the majority of oil produced in McElmo Creek.

3. Abundant secondary porosity results from meteoric solution of unstable carbonate minerals in shallow subtidal and peritidal facies that cap shallowing-upward cycles. Moldic porosity occurs throughout the Desert Creek and lower Ismay section, but is particularly common in cycle capping facies in highstand systems tracts. Moldic porosity is developed best on the carbonate platform below sequence boundaries in parasequence Layers IIA, IA, and LISA. Storage capacity is favorable, but permeability is low, generally less than 1 md. Facies composed dominantly of moldic porosity in the absence of significant primary porosity (interparticle or shelter) are poor reservoirs.

ACKNOWLEDGEMENTS

The authors wish to express thanks to the many people involved in the various aspects of the geologic assessment of the southern Paradox basin and the reservoir characterization study of the McElmo Creek Unit, Greater Aneth Field. Countless discussions and constructive comments have led to the improvement of ideas, which ultimately made their way into this paper. Major technical contributions have been made by the following people: John Armentrout, Don Best, Leslie Harman, Neil Humphreys, Jim Markello, Jose Olmos, Ed Shaw, and Jim Vanderhill. Additional thanks are due to Gordon Baker and Mike Croft for their administrative support, and to Bob Clarke for preparing the final page copy. The Dallas and Midland-based drafting and reprographics departments deserve special recognition for their assistance in the preparation of graphic displays.

---

*NOTE: Figures 17, 18, 19, 21, 23, 26, and 30 are oversized. These displays do not conform to SEPM publication guidelines. They are <u>NOT</u> included with the text. However, copies may be obtained by contacting:

Jim Weber
Mobil E&P Technical Center
P.O. Box 650232
Dallas, Texas 75265-0232.
   Telephone: (214)951-2691
   Fax: (214)951-2265;
   E-mail: ljweber@dal.mobil.com.

Photographic Plates

Plate 1

Selected features of Anhydrite and Cryptalgalaminite/Stromatolite Facies.

A. Massive anhydrite showing palmate growth. Subvertical white anhydrite columns separated by brown, laminated dolomudstone. Slab photo, MCU S-21, 5444', scale in inches.

B. Anhydrite crystal orientation is perpendicular to deposition. Dolomudstone lamina (L) parallels depositional fabric and serves as nucleation sites for vertically-directed crystal splays (AN). Photomicrograph, MCU V-06, 5864', plane-polarized light, scale in mm.

C. Laminae within the Cryptalgalaminite/Stromatolite Facies are smooth, wavy, and less visibly crinkly. Slab photo, MCU S-21, 5434', scale in inches.

D. Laminae are composed of intercalated dolomudstone (DM), siltstone (S), and anhydrite (AN). Photomicrograph, MCU S-21, 5436.5', cross-polarized light, scale in mm.

Plate 1

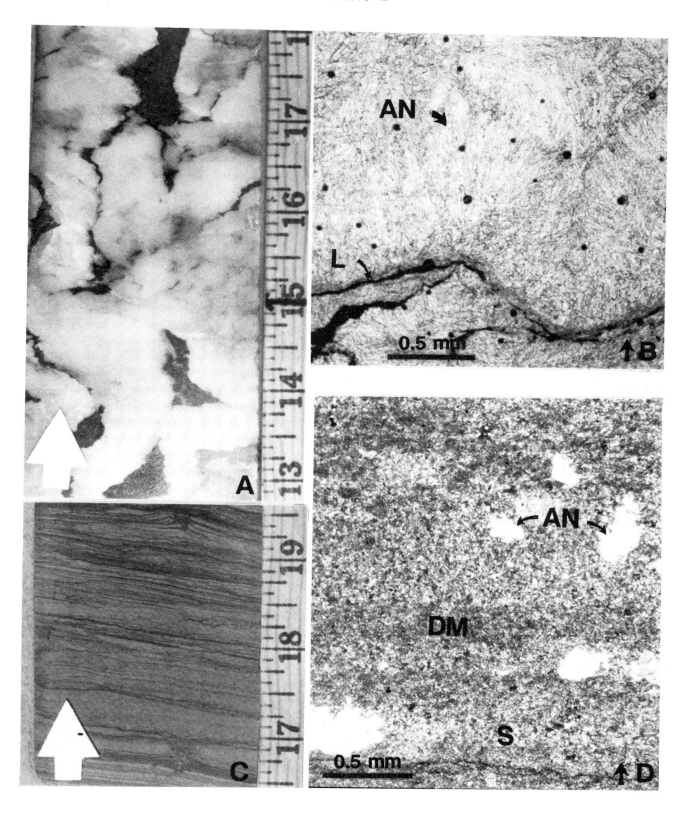

Plate 2

Selected features of Cryptalgalaminite/Stromatolite and Siltstone/Sandstone Facies.

A. Probable desiccation prism cracks disrupting cryptalgal laminae. Rock is pervasively dolomitized. Slab photo, MCU S-21, 5429', scale in inches.

B. Slab photo of tan-gray siltstone. Note fine-grained texture. Possible rhizoliths (R) (root casts) are indicated. Rhizoliths are diagnostic of subaerial exposure. Vadose silt is black and fills root casts. MCU Q-16, 5466', scale in inches.

C. Limestone peloids and debris (black) in siltstone matrix of the Siltstone/Sandstone Facies. Photomicrograph, MCU Q-16, 5466', plane-polarized light, scale in mm.

D. Quartz siltstone deposited in basin. Note dark gray to black color. Thin laminations are not readily apparent in slab photograph. MCU S-21, 5424', scale in inches.

Plate 2

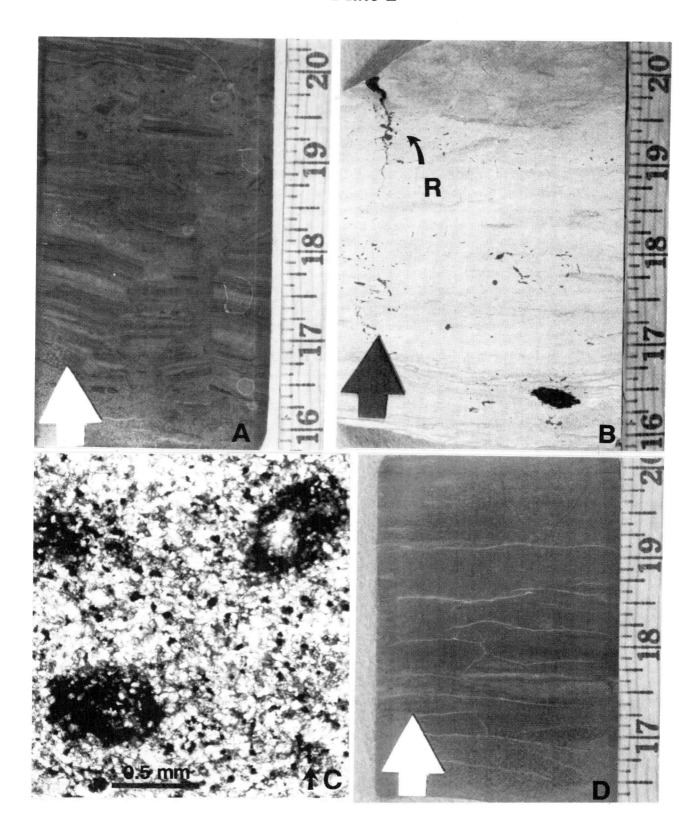

Plate 3

Selected features of Dolomudstone/Wackestone Facies.

A. Mud and spar filled fenestral fabric in dolomudstone. Slab photo, Aneth Unit K-231, 5711', scale in inches.

B. Clasts of dolomudstone in Dolomud-stone/Wackestone Facies are an indication of higher-energy conditions, possible tidal channels. Slab photograph, MCU K-231, 5798', scale in inches.

C. Millimeter-scale wispy laminations (upper half of photo) overlie nonlaminated zone. Note subvertical syndepositional cracks which emanate downward from base of laminated interval. Fractures are filled with anhydrite. Slab photo, Aneth Unit K-231, 5612', scale in inches.

D. Subvertical desiccation cracks/fissures located below thin laminated interval (top of photo). Anhydrite crystal laths replace dolostone matrix on right of slab photo. MCU K-231, 5603', scale in inches.

Plate 3

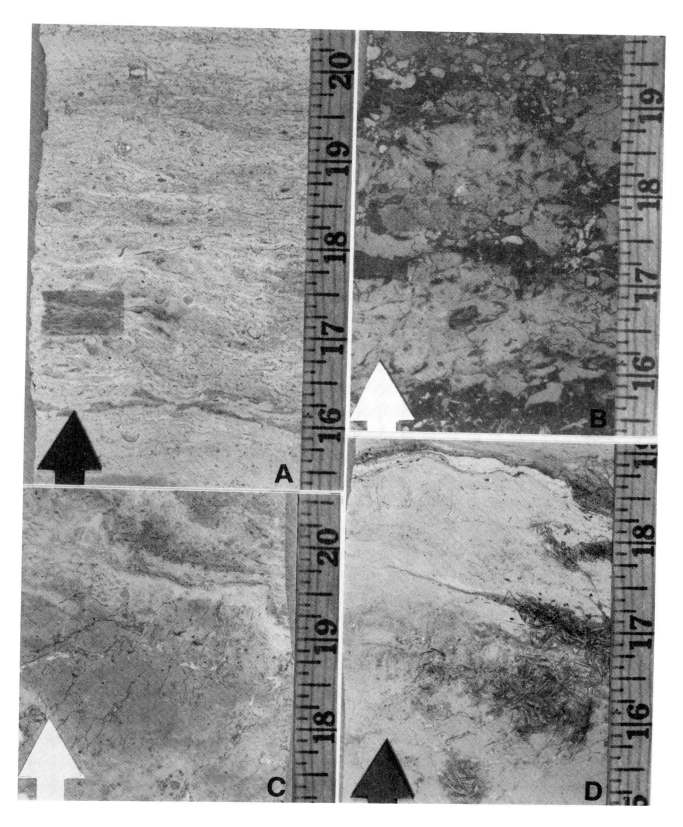

Plate 4

Selected features of Dolomudstone/Wackestone, Dolomitized Packstone/Grainstone, and Black Laminated Mudstone Facies.

A. Moldic and intercrystalline porosity (scattered throughout photomicrograph) in a fine-grained dolostone matrix. Leached allochems are often recognized as molluscs. Photomicrograph, MCU V-08, 5892', plane-polarized light, scale in mm.

B. Pervasively dolomitized ooids, peloids, and composite grains within the Dolomitized Packstone/Grainstone Facies. Photomicrograph, MCU U-18, 5560', cross-polarized light, scale in mm.

C. General character of Black Laminated Mudstone Facies. Dark color results from sulfides and residual organic matter. Photomicrograph, MCU O-16, 5709', plane-polarized light, scale in mm.

D. Gothic below (G) grading up into tidal flat dolomudstone (DM). Slab photo, MCU R-18, 5409', scale in inches.

Plate 4

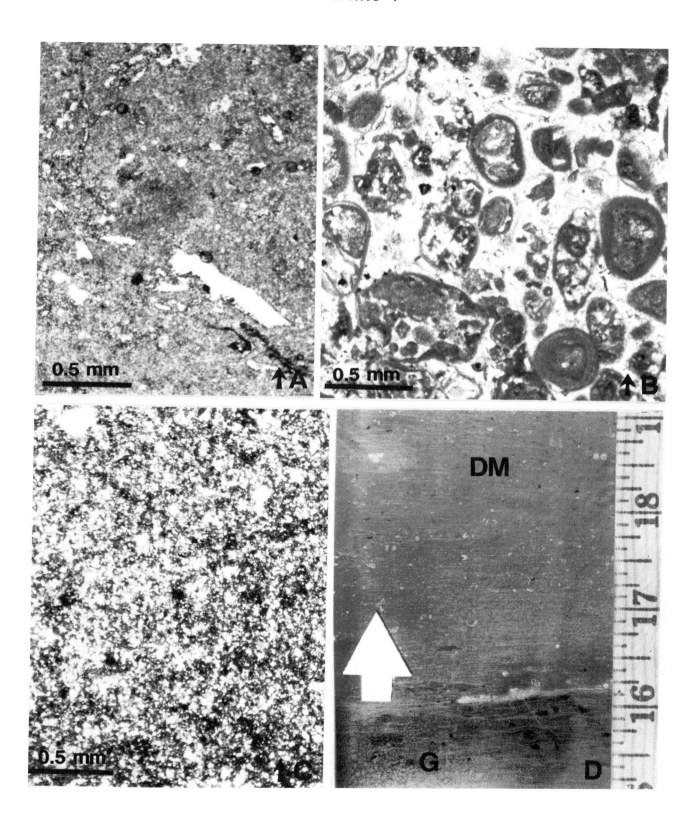

Plate 5

Selected features of Oolitic Grainstone and Non-Skeletal Packstone/Grainstone Facies.

A. Stabilized sand flat subfacies of the Oolitic Grainstone Facies. Note distinctive burrows. Oomoldic porosity is visible. Slab photo, Aneth Unit K-231, 5644', scale in inches.

B. Well-sorted, current stratified, and mud-free mobile sand belt of the Oolitic Grainstone Facies. Oomoldic porosity is readily apparent. Laminae are inclined and result from changes in cementation and grain size. Aneth Unit K-231, 5628', scale in inches.

C. Solution enhanced secondary porosity and silcrete deposits (SC) within a fine-grained grain-supported matrix of the Non-Skeletal/Skeletal Packstone/Grainstone Facies. Slab photo, Aneth Unit K-231, 5685, scale in inches.

D. Rhizoliths (root casts) in a fine-grained peloid grainstone. Irregular black clasts (right of up-arrow) are composed of calichified limestone. Dark gray siltstone fills root casts. Peloid and ooid grainstone clasts are observed at the top of the slab photo. Slab photo, MCU O-16, 5718', scale in inches.

Plate 5

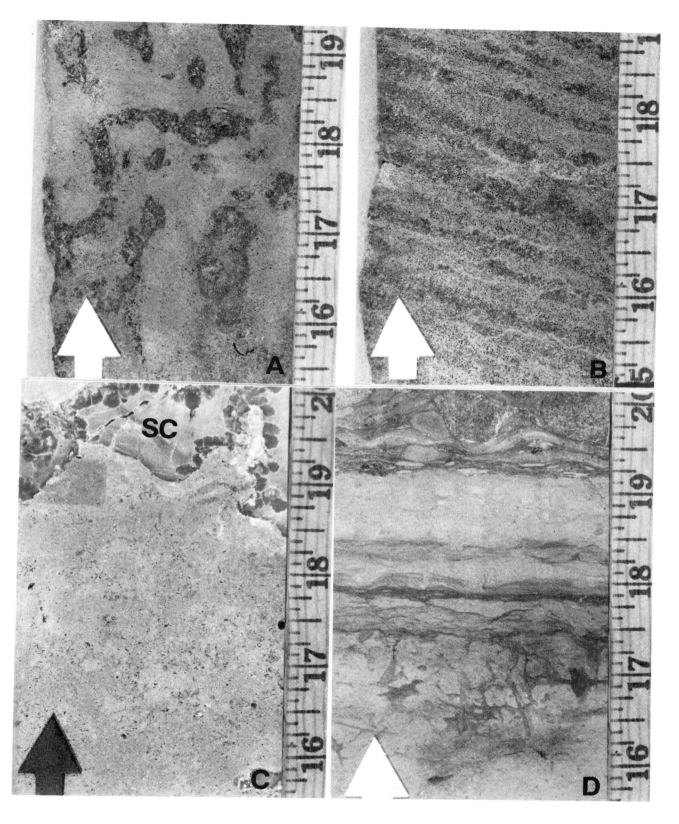

Plate 6

Selected features of Skeletal Packstone and Skeletal Wackestone Facies.

A. Skeletal packstone composed of crinoids, forams, and brachiopods. Slab photo, Aneth Unit K-231, 5723', scale in inches.

B. Packstone to poorly-washed grainstone composed entirely of encrusting forams. Photomicrograph, MCU R-19, 5731', cross-polarized light, scale in mm.

C. Skeletal wackestone composed dominantly of echinoderms. The dark gray to black color is a distinguishing characteristic of this subfacies. Slab photo, Aneth Unit K-231, 5618', scale in inches.

D. Skeletal wackestone composed of forams, molluscs, echinoderms, brachiopod spines, and finely comminuted skeletal debris. Photomicrograph, MCU S-21, 5517.4', plane-polarized light, scale in mm.

Plate 6

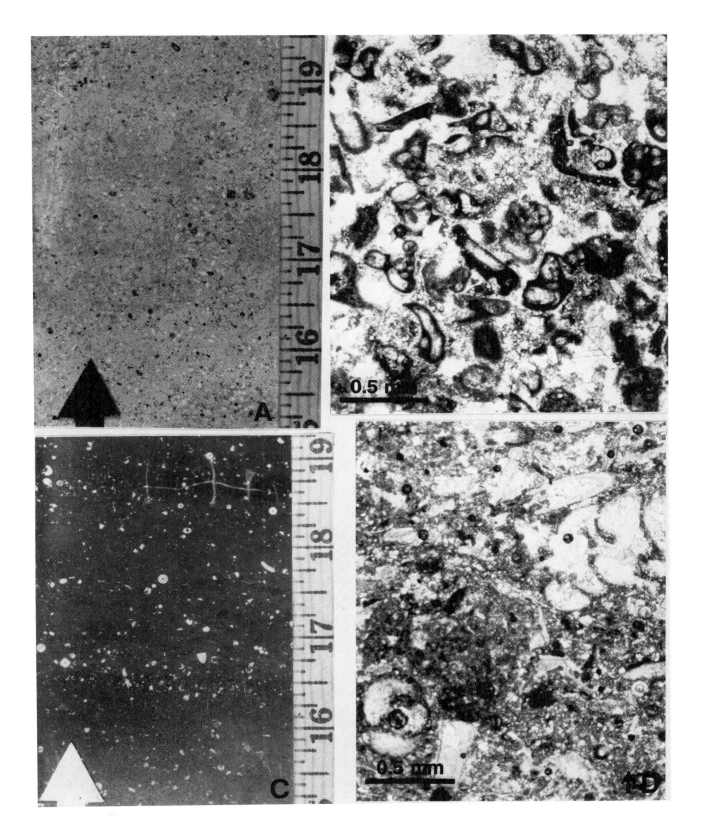

Plate 7

Selected features of Algal Facies.

A. General character of Phylloid Algal Bafflestone Facies showing abundant primary (shelter) porosity. Slab photo, MCU K-231, 5778', scale in inches.

B. Fibrous marine cement rimming phylloid algal plates. Note oil stained pores. Photomicrograph, MCU T-17, 5730', cross-polarized light, scale in mm.

C. General character of Phylloid Algal Packstone/Grainstone Facies. Interparticle porosity is formed by irregular packing of abraded and rounded algal plates. Slab photo, Aneth Unit K-231, 5752', scale in inches.

D. General character of phylloid algal wackestone Facies. White areas are blocky calcite cement. Molds and vugs are more abundant than interparticle porosity and result from leaching by meteoric water. Light gray areas are dolomitized. Slab photo, MCU Q-16, 5586', scale in inches.

Plate 7

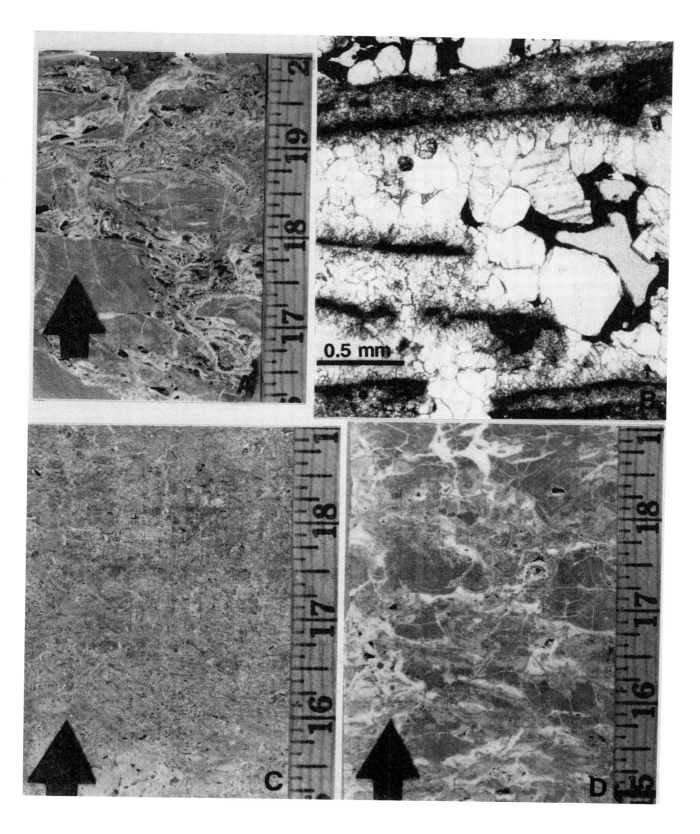

Plate 8

Selected features of Allochthonous Oolitic Grainstone and Allochthonous Peloidal/Skeletal Packstone Facies.

A. Fine-grained texture of Allochthonous Oolitic Grainstone Facies. Slab photo, MCU S-21, 5468', scale in inches.

B. Fine-grained ooids with well-preserved ooid cortices. Preserved ooid cortices indicate that meteoric water played minor role in altering these ooids. Also note well-preserved interparticle porosity. Photomicrograph, MCU S-21, 5461', cross-polarized light, scale in mm.

C. Very fine-grained texture of Allochthonous Peloidal/Skeletal Packstone Facies. Slab photograph, MCU S-21, 5485', scale in inches.

D. Very fine-grained peloids (P) in grain-supported rock texture. Photomicrograph, MCU V-08, 5875', cross-polarized light, scale in mm.

Plate 8

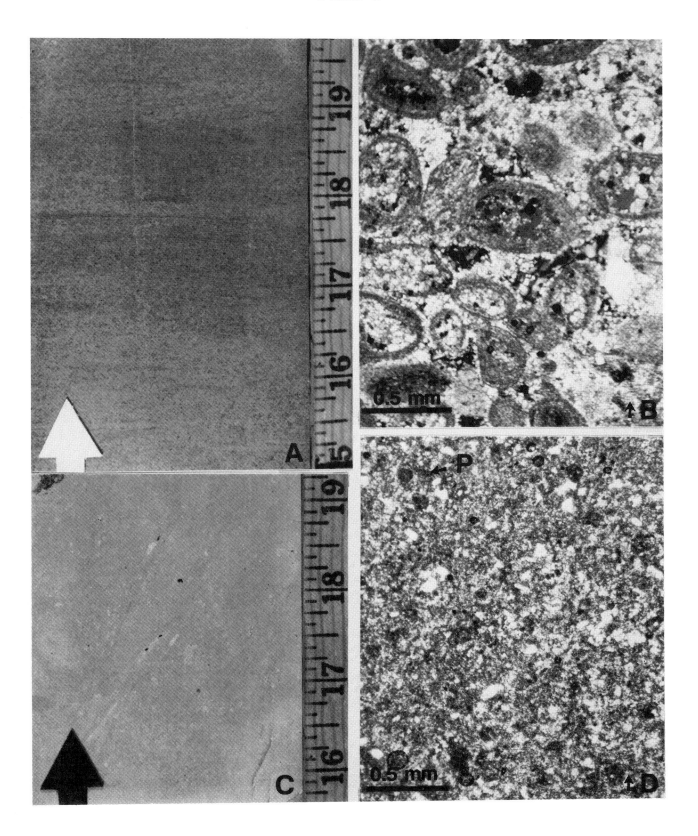

BIBLIOGRAPHY

Baars, D.L., 1962, Permian System of Colorado Plateau: American Association of Petroleum Geologists Bulletin, v. 46, p. 149-218.

Baars, D.L., 1966, Pre-Pennsylvanian Paleotectonics—Key to Basin Evolution and Petroleum Occurrences in the Paradox Basin, Utah and Colorado: American Association of Petroleum Geologists Bulletin, v. 50, p. 2082-2111.

Baars, D.L., 1979, The Permian System, in D.L. Baars, ed., Permianland: Four Corners Geological Society Guidebook, 9th Field Conference, p. 1-6.

Baars, D.L., Parker, J.W., and Chronic, J., 1967, Revised Stratigraphic Nomenclature of Pennsylvanian System, Paradox Basin: American Association of Petroleum Geologists Bulletin, v.51, p. 393-403.

Baars, D.L. and Stevenson, G.M., 1981, Tectonic Evolution of the Paradox Basin, Utah and Colorado in Wiegand, D.L., ed., Geology of the Paradox Basin: Rocky Mountain Association of Geologists 1981 Field Conference, p. 23-31.

Blakey, R.C. 1980, Pennsylvanian and Early Permian Paleogeography, Southern Colorado Plateau and Vicinity in Fouch, T.D., and Magathan, E.R. eds, Paleozoic Paleogeography of West-central United States: Society of Economic Paleontologists and Mineralogists, Rocky Mountain Section, West-central United States Paleogeography Symposium 1, p. 111-128.

Brown, L.F., Jr., 1969, Geometry and Distribution of Fluvial and Deltaic Sandstones (Pennsylvanian and Permian), North-Central Texas: Gulf Coast Association of Geological Society Transactions, v.19, p.23-47. Reprinted as Texas University Bureau Economic Geology Circular 69-4.

Brown, L.F., and Fisher, W.L., 1977, Seismic-Stratigraphic Interpretation of Depositional Systems: Examples from Brazil Rift and Pull-Apart Basins, in C.E. Payton, ed., Seismic Stratigraphy-Applications to Hydrocarbon Exploration: American Association of Petroleum Geologists Memoir 26, p. 213-248.

Byers, C.W., 1977, Biofacies Patterns in Euxinic Basins: a General Mode, in Cook, H.E. and Enos, P., eds., Deep-water Carbonate Environments: Society of Economic Paleontologists and Mineralogists Special Publication no. 25, p. 5-18.

Campbell, J.A., 1979, Lower Permian Depositional System, Northern Uncompahgre Basin in D.L. Baars, ed., Permianland: Four Corners Geological Society 9th Field Conference Guidebook, p. 13-21.

Campbell, J.A., 1980, Lower Permian Depositional Systems, "Uncompahgre" Basin, Eastern Utah and Southwestern Colorado (abstract): American Association of Petroleum Geologists Bulletin, v. 64, p. 686.

Choquette, P.W. 1983, Platy Algal Reef Mounds, Paradox Basin in Scholle, P.A., Bebout, D.G., and Moore, C.H., eds., Carbonate Depositional Environments: American Association of Petroleum Geologists Memoir no. 33, p. 454-462.

Choquette, P.W. and Traut, J.D., 1963, Pennsylvanian Carbonate Reservoirs, Ismay Field, Utah and Colorado, in Bass, R.O., ed., Shelf Carbonates of the Paradox Basin: Four Corners Geological Society 4th Field Conference Guidebook, P. 157-184.

Clair, J.R., 1952, Paleozoic Rocks of the Southern Paradox Basin, in Geological Symposium of the Four Corners Region: Four Corners Geological Society, p. 36-39.

Clair, J.R., 1958, Subsurface stratigraphy of the Pennsylvanian of the Paradox Basin, in Symposium on Pennsylvanian Rocks of Colorado and Adjacent Areas: Rocky Mountain Association of Geologists, Denver, Colorado, p. 31-47.

Clem, K.M. and Brown, K.W., 1984, Petroleum Resources of the Paradox Basin: Utah Geological and Mineral Survey, Bulletin 119, pp. 74-75.

Copeland, B.J., 1967, Environmental Characteristics of Hypersaline Lagoons: Contributions to Marine Science, v. 12, p. 207-218.

Crevello, P., Sarg, J.F., Read, J.F., and Wilson, J.L., eds, 1989, Controls on Carbonate Platform to Basin Development: Society of Economic Paleontologists and Mineralogists Special Publication no. 44, 405 p.

Dawson, W.C., 1988, Ismay Reservoirs, Paradox Basin; Diagenesis and Porosity Development: Rocky Mountain Association of Geologists 1988 Carbonate Symposium, p. 163-174.

Doelling, H.H., 1983, Observations on Paradox Basin Salt Anticlines, in Averett, W.R., ed., Northern Paradox Basin—Uncompahgre Uplift: Grand Junction Geological Society Field Trip, p. 81-90.

Donovan, D.T., and Jones, E., 1979, Causes of World-Wide Changes in Sea Level: Journal of the Geological Society of London, v. 136, p. 187-192.

Driese, S.G., and Dott, R.H. 1984, Model for Sandstone-Carbonate "Cyclothems" Based on Upper Member of Morgan Formation (Middle Pennsylvanian) of Northern Utah and Colorado: American Association of Petroleum Geologists Bulletin, v. 68, p. 574-597.

Elias, G.K., 1963, Habitat of Pennsylvanian Algal Bioherms, Four Corners Area, in Bass, R.O., ed., Shelf Carbonates of the Paradox Basin: Four Corners Geological Society 4th Field Conference Guidebook, p. 185-203.

Elston, D.P., Shoemaker, E.M., and Landis, E.R., 1962, Uncompahgre Front and Salt Anticline Region of Paradox Basin, Colorado and Utah: American Association of Petroleum Geologists Bulletin, v. 46, p. 1857-1878.

Embry, A.E. and Klovan, J.E., 1971, A Late Devonian Reef Tract on Northeastern Banks Island, Northwest Territories: Bulletin of Canadian Petroleum Geology, v. 19, p. 730-781.

Fetzner, R.W., 1960, Pennsylvanian Paleotectonics of Colorado Plateau: American Association of Petroleum Geologists Bulletin, v. 44, p. 1371-1413.

Frahme, C.W., and Vaughn, E.B., 1983, Paleozoic Geology and Seismic Stratigraphy of the Northern Uncompahgre Front, Grand County, Utah in J.D. Lowell, ed., Rocky Mountain Foreland Basins and Uplifts: Rocky Mountain Association of Geologists, Denver, Colorado, p. 201-211.

Goldhammer, R. K., Oswald, E. J. and Dunn, P. A., 1991, The Hierarchy of Stratigraphic Forcing: an Example from Middle Pennsylvanian Shelf Carbonates of the Paradox Basin, in Franseen, E. K., and others, eds., Sedimentary Modelling: Computer Simulations and Methods for Improved Parameter Definition: Kansas Geological Survey, Bulletin 233, p. 361-413.

Gray, R.S., 1967, Cache Field - A Pennsylvanian Algal Reservoir in Southwestern Colorado: American Association Petroleum Geologists Bulletin, v. 51, p. 1959-1978.

Haq, B.U., Hardenbol, J., and Vail, P.R., 1987, Chronology of Fluctuating Sea Levels Since the Triassic: Science, v.235, p.1,156-1,166.

Harland, W.B., Armstrong, R.L., Cox, A.V., Craig, L.E., Smith, A.G., Smith, D.G., 1989, A Geologic Time Scale: Cambridge University Press, New York, 128p.

Harris, P. M., 1979, Facies Anatomy of a Bahamian Ooid Shoal: Sedimenta VII, The Comparative Sedimentology Laboratory, The University of Miami, 163 p.

Heckel, P.H., 1977, Origin of Phosphatic Black Shale Facies in Pennsylvanian Cyclothems of Mid-Continents North America: American Association of Petroleum Geologists Bulletin, v. 61, p. 1045-1068.

Heckel, P.H., 1980, Paleogeography of Eustatic Model for Deposition of Midcontinent Upper Pennsylvanian Cyclothems, in T.D. Fouch, and E.R. Magathan, eds., Rocky Mountain Section, Society of Economic Paleontologists and Mineralogists West-Central U.S. Paleogeography Symposium I, Paleozoic Paleogeography of West-Central United States, p. 197-215.

Heckel, P. H., 1986, Sea-level Curve for Pennsylvanian Eustatic Marine Transgressive-Regressive Despositional Cycles along Midcontinent Outcrop Belt, North America: Geology, v. 14, p. 330-334.

Herman, G., and Sharps, S.L., 1956, Pennsylvanian and Permian Stratigraphy of the Paradox Salt Embayment, in Geology and Economic Deposits of East Central Utah: Intermountain Association of Petroleum Geologists, Seventh Annual Field Conference, p. 77-84.

Herrod, W.H. and Gardner, P.S., 1988, Upper Ismay Reservoir at Tin Cup Mesa Field: Rocky Mountain Association of Geologists 1988 Carbonate Symposium, p.175-192.

Herrod, W. H., Roylance, M. H., and Strathouse, E. C., 1985, Pennsylvanian Phylloid-Algal Mound Production at Tin Cup Mesa Field, Paradox Basin, Utah, in Longman, M. W., and others, eds., Rocky Mountain Carbonate Reservoirs: A Core Workshop: Society of Economic Paleontologists and Mineralogists Core Workshop No. 7, Golden, Colorado, p. 409-445.

Hite, R.J., 1960, Stratigraphy of the Saline Facies of the Paradox Member of the Hermosa Formation of Southeastern Utah and Southwestern Colorado, in Geology of the Paradox Basin Fold and Fault Belt: Four Corners Geological Society, 3rd Annual Field Conference, p. 86-89.

Hite, R.J., 1970, Shelf Carbonate Sedimentation Controlled by Salinity in the Paradox Basin, Southeast Utah, in Ron J.L. and Dellwig, L.F., eds., Third Symposium on Salt: Northern Ohio Geologic Society, v. 1, p. 48-66.

Hite, R.J. and Buckner, D.H., 1981, Stratigraphic Correlations, Facies Concepts, and Cyclicity in Pennsylvanian Rocks of the Paradox Basin, in Wiegand, D.L., ed., Geology of the Paradox Basin: Rocky Mountain Association of Geologists 1981 Field Conference, p. 147-159.

James, N. P., 1972, Holocene and Pleistocene Calcareous Crust (Caliche) Profiles: Criteria for Subaerial Exposure: Journal of Sedimentary Petrology, v. 42, p. 817-836.

Jervey, M.T., 1988, Quantitative Geological Modeling of Siliciclastic Rock Sequences and Their Seismic Expression; in, Sea-Level Changes: An Integrated Approach, Wilgus, C.K., et al., eds., Society of Economic Paleontologists and Mineralogists Special Publication No. 42, p.47-70.

Kendall, G. St. D., and Schlager, W., 1981, Carbonates and Relative Changes in Sea Level: Marine Geology v. 44, p. 181-212.

Kinsman, D.J., Boardman, M., and Borcsik, M., 1973, An Experimental Determination of the Solubility of Oxygen in Marine Brines, in Fourth Symposium on Salt: Northern Ohio Geological Society, v.1, p. 323-327.

Kirkland, D.W. and Evans, R., 1981, Source-Rock Potential of Evaporite Environment: American Association of Petroleum Geologists Bulletin, v. 65, p. 181-189.

Kluth, C.F. and P.J. Coney, 1981, Plate Tectonics of the Ancestral Rocky Mountains: Geology, v. 9, p. 10-15.

Loope, D.A., 1984, Eolian Origin of Upper Paleozoic Sandstones, Southeastern Utah: Journal of Sedimentary Petrology, v. 54, p. 563-580.

Mack, G.H., 1977, Depositional Environments of the Cutler-Cedar Mesa Facies Transition (Permian) near Moab, Utah: Mountain Geologist, v. 14, p. 53-68.

Mack, G.H., 1979, Littoral Marine Depositional Model for the Cedar Mesa Sandstone (Permian), Canyonlands National Park, Utah in D.L. Baars, ed., Permianland: Four Corners Geological Society 9th Field Conference Guidebook, p. 13-21.

Mack, G.H., and K.A. Rasmussen, 1984, Alluvial-Fan Sedimentation of the Cutler Formation (Permo-Pennsylvanian) near Gateway, Colorado: Geological Society of America, v. 95, p. 109-116.

Mallory, W.W. 1958, Pennsylvanian Coarse Arkosic Red Beds and Associated Mountains in Colorado, in Symposium on Pennsylvanian Rocks of Colorado and Adjacent Areas: Rocky Mountain Association of Geologists, Denver, Colorado, p. 17-20.

Mallory, W.W., 1972, Pennsylvanian Arkose and the Ancestral Rocky Mountains, in Geologic Atlas of the Rocky Mountain Region: Rocky Mountain Association of Geologists, Denver, Colorado, p. 131-132.

Merrill, W.M., and R.M. Winar, 1958, Molas and Associated Formations in San Juan Basin, Needle Mountains area, Southwestern Colorado: American Association of Petroleum Geologists Bulletin, v. 42, p. 2107-2132.

Mitchum, R.M., 1977, Seismic Stratigraphy and Global Changes of Sea Level, Part 1: Glossary of Terms used in Seismic Stratigraphy, in C.E. Payton, ed., Seismic Stratigraphy-Applications to Hydrocarbon Exploration: American Association of Petroleum Geologists Memoir 26, p. 205-212.

Mitchum, R.M., and Van Wagoner, J.C., 1991, High-Frequency Sequences and Their Stacking Patterns: Sequence-Stratigraphic Evidence of High-Frequency Eustatic Cycles: Sedimentary Geology, v.70, p.131-160.

Molenar, C.M., 1981, Mesozoic Stratigraphy of the Paradox Basin—An Overview, in D.L. Weigand, ed., Geology of the Paradox Basin: Rocky Mountain Association of Geologists, 1981 Field Conference, p. 119-127.

Odin, G.S., and Gale, N.H., 1982, Mise a jour d'echelles des temps caledoiens et hercyniens: Academie des Sciences, Comptes Rendus, v.294, series 2, p.453-456.

Ohlen, H.R., and McIntyre, L.B., 1965, Stratigraphy and Tectonic Features of Paradox Basin, Four Corners Area: American Association of Petroleum Geologists Bulletin, v. 49, p. 2020-2040.

O'Sullivan, R.B., 1965, Geology of the Cedar Mesa-Boundary Butte Area, San Juan County, Utah: United States Geological Survey Bulletin 1186, p. 128.

Parrish, J.T., 1982, Upwelling and Petroleum Source Beds, with Reference to Paleozoic: American Association of Petroleum Geologists Bulletin, v. 66, p. 750-774.

Peterson, J.A., 1966, Stratigraphic vs. Structural Controls on Carbonate-Mound Hydrocarbon Accumulation, Aneth Area, Paradox Basin: American Association of Petroleum Geologists Bulletin, v. 50, p. 2068-2081.

Peterson, J.A., 1992, Aneth Field, USA, Paradox Basin, Utah; in Stratigraphic Traps III, Foster, N.H. and Beaumont, E.A., eds.: Treatise of Petroleum Geology, Atlas of Oil and Gas Fields, p. 41-82.

Peterson, J.A., and Hite, R.J., 1969, Pennsylvanian Evaporate Carbonate Cycles and Their Relation to Petroleum Occurrence: American Association of Petroleum Geologists Bulletin, v. 53, p. 884-908.

Patterson, R.J. and Kinsman, D.J.J., 1982, Formation of Diagenetic Dolomite in Coastal Sabkha along Arabian (Persian) Gulf; American Association of Petroleum Geologists Bulletin, v. 66, p. 28-43.

Peterson, J.A., Loleit, A.J., Spencer, C.W., and Ullrich, R.A., 1965, Sedimentary History and Economic Geology of San Juan Basin: American Association of Petroleum Geologists Bulletin, v. 49, p. 2076-2119.

Peterson, J.A., and Ohlen, H.R., 1963, Pennsylvanian Shelf Carbonates, Paradox Basin, in R.O. Bass, ed., Shelf Carbonates of the Paradox Basin: Four Corners Geological Society 4th Field Conference Symposium, p. 65-79.

Posamentier, H.W., Jervey, M.T., and Vail, P.R., 1988, Eustatic Controls on Clastic Deposition I - Conceptual Framework; in, Sea-Level Changes: An Integrated Approach, Wilgus, C.K., et al., eds.: Society of Economic Paleontologists and Mineralogists Special Publication No. 42, p.109-124.

Pray, L.C. and Wray, J.L., 1963, Porous Algal Facies (Pennsylvanian) Honaker Trail, San Juan Canyon, Utah, in R.O. Bass, ed., Shelf Carbonates of the Paradox Basin: Four Corners Geological Society, 4th Field Conference Guidebook, p. 204-234.

Ross, C.A. and Ross, J.R.P., 1985, Late Paleozoic Depositional Sequences are Synchronous and Worldwide: Geology, v. 13, p. 194-197.

Ross, C.A., and Ross, J.R.P., 1987a, Late Paleozoic Sea Levels and Depositional Sequences, in Ross, C.A., and Haman, D., eds., Timing and Depositional History of Eustatic Sequences: Constraints on Seismic Stratigraphy: Cushman Foundation for Foraminiferal Research, Special Publication 24, p.137-149.

Ross, C.A., and Ross, J.R.P., 1987b, Biostratigraphic Zonation of Late Paleozoic Depositional Sequences, in Ross, C.A., and Haman, D., eds., Timing and Depositional History of Eustatic Sequences: Constraints on Seismic Stratigraphy: Cushman Foundation for Foraminiferal Research, Special Publication 24, p.151-168.

Rossinsky, V. and Wanless, H. R., 1992, Topographic and Vegetative Controls on Calcrete Formation, Turks and Caicos Islands, British West Indies: Journal of Sedimentary Petrology, v. 62, p. 84-98.

Roylance, M. H., 1984, Depositional and Diagenetic Control of Petroleum Entrapment in the Desert Creek Interval, Paradox Formation, Southeastern Utah and Southwestern Colorado: Unpublished MS. Thesis, University of Kansas.

Sarg, J.F., 1988, Carbonate Sequence stratigraphy, in Wilgus, C.K., et al., eds., Sea-Level Changes – An Integrated Approach, Society of Economic Paleontologists and Mineralogists Special Publication No. 42, p.155-181.

Schlager, W., 1981, The Paradox of Drowned Reefs and Carbonate Platforms: Geological Society of American Bulletin, v. 92, p. 197-211.

Sloss, L.L., 1963, Sequences in the Cratonic Interior of North America: Geological Society of America Bulletin, v. 74, p.93-114.

Sloss, L.L., Krumbein, W.C., and Dapples, E.C., 1949, Integrated Facies Analysis; in Longwell, C.R., chairman, Sedimentary Facies in Geologic History, Geological Society of America, Memoir 39, p.91-124.

Szabo, E., and Wengerd, S.A., 1975, Stratigraphy and Tectogenesis of the Paradox Basin; Canyonlands: Four Corners Geological Society Field Conference and Guide No. 8, p. 193-210.

Vail, P. R., 1987, Seismic Stratigraphy Interpretation Procedure; in, Atlas of Seismic Stratigraphy, Bally, A. W., ed.: American Association of Petroleum Geologists, Studies in Geology 27, v. 1, p. 1-11.

Vail, P.R., Audemard, F., Bowman, S.A., Eisner, P.N., and Perez-Cruz, C., 1991, The Stratigraphic Signatures of Tectonics, Eustasy and Sedimentology - an Overview; in Einsele, G., et al., eds., Cycles and Events in Stratigraphy: Springer-Verlag, Berlin-Heidelberg, p.617-661.

Vail, P.R., Mitchum, R.M., Jr., and Thompson, S. III, 1977, Seismic Stratigraphy and Global Changes of Sea Level, in Payton, C.E., ed., Seismic Stratigraphy - Applications to Hydrocarbon Exploration: American Association of Petroleum Geologists, Memoir 26, p. 82-97.

Van Eysinga, F.W.B., 1975, Geologic Time Table: Elsevier Scientific Publication Company, Amsterdam, 1p.

Van Wagoner, J.C., Mitchum, R.M., Campion, K.M., and Rahmanian, V.D., 1990, Siliciclastic Stratigraphy in Well Logs, Cores, and Outcrops; Concepts for High-Resolution Correlation of Time and Facies: American Association of Petroleum Geologists, Methods in Exploration Services, No. 7, 55 p.

Van Wagoner, J.C., Posamentier, H.W., Mitchum, R.M., Vail, P.R., Sarg, J.F., Loutit, T.S., and Hardenbol, J., 1988, An Overview of the Fundamentals of Sequence Stratigraphy and Key Definitions; in Wilgus, C.K., et al., eds., Sea-Level Changes - An Integrated Approach: Society of Economic Paleontologists and Mineralogists Special Publication No. 42, p.39-46.

Warren, J.K. and Kendall, C.G.St.C., 1985, Comparison of Sequences Formed in Marine Sabkha (subaerial) and Salina (subaqueous) Settings-Modern and Ancient: American Association of Petroleum Geologists Bulletin, v. 69, p. 1013-1023.

Weber, L.J., 1992, Geologic Description and Aspects of Reservoir Description (Desmoinesian) carbonates, Aneth field, McElmo Creek Unit, SE Utah: Mobil Internal Report, 290 p.

Weber, L.J. and Wright, F.M., 1991, Greater Aneth Oil Field and Honaker Trail Stratigraphic Section, Paradox Basin Field Seminar, SE Utah, Fieldtrip Guidebook: Mobil-Midland Internal Report, 125 p.

Weber, L.J., Wright, F.M., Sarg, J.F., Shaw, E., Harman, L.P., Vanderhill, J.P., and Best, D.A., 1995, Reservoir Delineation and Performance: Application of Sequence Stratigraphy and Integration of Petrophysics and Engineering Data, Aneth Field, Southeast Utah, U.S.A.; in Stoudt, E.L. and Harris, P.M., eds., Hydrocarbon Reservoir Characterization: Geologic Framework and Flow Unit Modeling: Society of Sedimentary Geology (SEPM), Short Course No. 34, p. 1-30.

Wengerd, S.A., 1950, Photogeologic Characteristics of Paleozoic Rocks on the Monument Upwarp, Utah: Photogram. Engineering, v. 16, no. 5, p. 770-781.

Wengerd, S.A., 1951, Reef limestones of Hermosa Formation, San Juan Canyon, Utah: American Association of Petroleum Geologists Bulletin, v. 35, p. 1038-1051.

Wengerd, S.A., 1955, Biohermal Reef Trends in Pennsylvanian Strata of San Juan Canyon, Utah in Geology of Parts of Paradox, San Juan and Black Mesa Basins: Four Corners Geological Society 1st Annual Field Conference.

Wengerd, S.A., 1958, Pennsylvania Stratigraphy, Southwest Shelf, Paradox Basin, in Intermountain Association of Petroleum Geologists, 9th Annual Field Conference Guidebook, Geology of the Paradox basin, p. 109-134.

Wengerd, S.A., 1962, Pennsylvanian Sedimentation in the Paradox Basin, Four Corners Region, in Pennsylvanian System in the United States...A Symposium: American Association of Petroleum Geologists, p. 264-330.

Wengerd, S.A., 1963, Stratigraphic Section At Honaker Trail, San Juan Canyon, San Juan County, Utah: Four Corners Geological Society, pp. 235-243.

Wengerd, S.A., 1973, Regional Stratigraphic Control of the Search for Pennsylvanian Petroleum, Southern Monument Upwarp, Southeastern Utah, in H.L. James, ed., Guidebook of Monument Valley and Vicinity, Arizona and Utah: New Mexico Geological Society, 24th Field Conference, p. 122-138.

Wengerd, S.A., and Matheny, M.L. 1958, Pennsylvanian System of the Four Corners Region: American Association of Petroleum Geologists Bulletin, v. 42, no. 9, p. 2048-2106.

Wengerd, S.A., and Strickland, J.W., 1954 Pennsylvanian Stratigraphy of the Paradox Salt Basin, Four Corners Region, Colorado and Utah: American Association Petroleum Geologists Bulletin, v. 38, no. 10, p. 2157-2199.

Wheeler, H.E., 1958, Time-Stratigraphy: American Association Petroleum Geologists Bulletin, V. 42, p.1047-1063.

Wilson, J.A., 1975, Carbonate Facies in Geologic History: Springer-Verlag, New York, 471 p.

Wright, V.P., Platt, N. H., and Wimbledon, W. A., 1988, Biogenic Laminar Calcretes: Evidence of Calcified Root-mat Horizons in Paleosols: Sedimentology, v. 35, p. 603-620.